T0286430

Underwater Acoustics

Underwater Acoustics

Edited by **Sonny Lin**

New Jersey

Published by Clanrye International,
55 Van Reypen Street,
Jersey City, NJ 07306, USA
www.clanryeinternational.com

Underwater Acoustics
Edited by Sonny Lin

International Standard Book Number: 978-1-63240-508-1 (Hardback)

Printed in the United States of America.

Contents

Preface

This book provides up-to-date information as well as introduction to underwater acoustics, which is described as the analysis of the propagation of sound in water and the interplay of the mechanical waves that constitute sound with the water and its boundaries. A wide range of topics are encompassed in this book like localization of buried objects in sediment with the help of high resolution array processing techniques, underwater acoustic source localization, adaptive strategy for underwater acoustic communication, etc. Researchers and scientists from across the world have contributed valuable data and information in this all-inclusive book. The aim of this elucidative book is to serve as a useful source of reference for readers including researchers, students and even scientists who are interested in acquiring knowledge regarding this field.

The information contained in this book is the result of intensive hard work done by researchers in this field. All due efforts have been made to make this book serve as a complete guiding source for students and researchers. The topics in this book have been comprehensively explained to help readers understand the growing trends in the field.

I would like to thank the entire group of writers who made sincere efforts in this book and my family who supported me in my efforts of working on this book. I take this opportunity to thank all those who have been a guiding force throughout my life.

Editor

1

A Novel Bio-Inspired Acoustic Ranging Approach for a Better Resolution Achievement

Said Assous[1], Mike Lovell[1], Laurie Linnett[2], David Gunn[3], Peter Jackson[3] and John Rees[3]
[1]*Ultrasound Research Laboratory, University of Leicester*
[2]*Fortkey Ltd*
[3]*Ultrasound Research Laboratory, British Geological Survey*
United Kingdom

1. Introduction

Bat and dolphin use sound to survive and have greatly superior capabilities to current technology with regard to resolution, object identification and material characterisation. Some bats can resolve some acoustic pulses thousands of times more efficiently than current technology (Thomas & Moss, 2004). Dolphins are capable of discriminating different materials based on acoustic energy, again significantly out-performing current detection systems. Not only are these animals supreme in their detection and discrimination capabilities, they also demonstrate excellent acoustic focusing characteristics - both in transmission and reception. If it could approach the efficiencies of bat and cetacean systems, the enormous potential for acoustic engineering, has been widely recognised. Whilst some elements of animal systems have been applied successfully in engineered systems, the latter have come nowhere near the capabilities of the natural world. Recognizing that engineered acoustic systems that emulate bat and cetacean systems have enormous potential, we present in this chapter a breakthrough in high-resolution acoustic imaging and physical characterization based on bio-inspired time delay estimation approach. A critical limitation that is inherent to all current acoustic technologies, namely that detail, or resolution, is compromised by the total energy of the system. Instead of using higher energy signals, resulting in poorer sound quality, random noise and distortion, they intend to use specifically designed adaptable lower energy 'intelligent' signals. There are around 1000 species of bats alive in the world today. These are broken down into the megabats, which include the large fruit bats, and the microbats, which cover a range of species, both small and large, which eat insects, fruit, nectar, fish, and occasionally other bats. With the exception of one genus, none of the megabats use echolocation, while all of the microbats do. Echolocation is the process by which the bat sends out a brief ultrasonic sound pulse and then waits to hear if there is an echo. By knowing the time of flight of the sound pulse, the bat can work out the distance to the target; either prey or an obstacle. That much is easy, and this type of technology has long been adopted by engineers to sense objects at a distance using sound, and to work out how far away they are. However, bats can do much more than this, but the extent of their abilities to sense the world around them is largely unknown, and the research often contradictory. Some experiments have shown that bats can time pulses, and hence work out the distance to objects with far greater accuracy than is currently possible, even to engineers. Sonar is a relatively recent invention by man for locating objects under water using sound waves. However, locating objects in water and air has evolved in the biological world to a much higher level of sophistication. Echolocation, often called biosonar, is used by bats and cetaceans (whales, manatees, dolphins etc.) using sound waves at ultrasonic frequencies (above 20 kHz). Based on the frequencies in the emitted pulses, some bats can resolve targets many times smaller than should be possible. They are clearly processing the sound differently to current sonar technology. Dolphins are capable of discriminating different materials based on acoustic energy, again significantly out-performing current detection systems. A complete review of this capabilities can be found in (Whitlow, 1993). Not only are these animals supreme in their detection and discrimination capabilities, they also demonstrate excellent acoustic focusing characteristics - both in transmission and reception. What we can gain from these animals is how to learn to see using sound. This approach may not lead us down the traditional route of signal processing in acoustic, but it may let us explore different ways of analyzing information, in a sense, to ask the right question rather than look for the right answer.

This chapter presents a bio-inspired approach for ranging based on the use of phase measurement to estimate distance (or time delay). We will introduce the technique with examples done for sound in air than some experiments for validation are done in tank water. The motivation for this comes from the fact that bats have been shown to have very good resolution with regard to target detection when searching during flight. Jim Simmons (Whitlow & Simmons, 2007) has estimated for bats using a pulse signal with a centre frequency of about 80 kHz (bandwidth 40 kHz) can have a pulse/echo resolution of distance in air approaching a few microns. For this frequency, the wavelength (λ) of sound in air is about 4 mm, and so using the half wavelength ($\lambda/2$) as the guide for resolution we see that this is about 200 times less than that achieved by the bat. We demonstrate in this chapter how we have been inspired from bat and its used signal (chirp) to infer a better resolution for distance measurement by looking to the phase difference of two frequency components.

2. Time delay and distance measurement using conventional approaches

Considering a constant speed of sound in a medium, any improvement in distance measurement based on acoustic techniques will rely on the accuracy of the time delay or the time-of-flight measurement. The time delay estimation is also a fundamental step in source localization or beamforming applications. It has attracted considerable research attention over the past few decades in different technologies including radar, sonar, seismology, geophysics, ultrasonics, communication and medical ultrasound imaging. Various techniques are reported in the literature (Knapp & Carter, 1976; Carter, 1979; 1987; Boucher & Hassab, 1981; Chen et al., 2004) and a complete review can be found in (Chen et al., 2006). Chen *et.al* in their review consider critical techniques, limitations and recent advances that have significantly improved the performance of time-delay estimation in adverse environments. They classify these techniques into two broad categories: correlator-based approaches and system-identification-based techniques. Both categories can be implemented using two or more sensors; in general, more sensors lead to increase robustness due to greater redundancy. When the time delay is not an integral multiple of the sampling rate, however, it is necessary to either increase the sampling rate or use interpolation both having significant limitations. Interpolating by using a parabolic fit to the peak usually yields to a biased estimate of the time delay, with both the bias and variance of the estimate dependent on the location of the delay between samples, SNR, signal and noise bandwidths, and the prefilter or window used in the generalized correlator. Increasing the sampling rate is not desirable for practical implementation, since sampling at lower rates is suitable for analog-to-digital converters (ADCs) that are more precise and have a lower power consumption. In addition, keeping the sampling rate low can reduce the load on both hardware and further digital processing units.

In this chapter, we present a new phase-based approach to estimate the time-of-flight, using only the received signal phase information without need to a reference signal as it is the case for other alternative phase-based approaches often relying on a reference signal provided by a coherent local oscillator (Belostotski et al., 2001) to count for the number of cycles taking the signal to travel a distance. Ambiguities in such phase measurement due to the inability to count integer number of cycles (wavelengths) are resolved using the Chinese Remainder Theorem (CRT) taken from number theory, where wavelength selection is based on pair-wise relatively prime wavelengths (Belostotski et al., 2001; Towers et al., 2004). However, the CRT is not robust enough in the sense that a small errors in its remainders may induce a large error in the determined integer by the CRT. CRT with remainder errors has been investigated in the literature (Xiang et al., 2005; Goldreich et al., 2000). Another phase-based measurement approach adopted to ensure accurate positioning of commercial robots, uses two or more frequencies in a decade scale in the transmitted signal. In this, the phase shift of the received signal with respect to the transmitted signal is exploited for ranging (Lee et al., 1989; Yang et al., 1994). However, this approach is valid only when the maximum path-length/displacement is less than one wavelength, otherwise a phase ambiguity will appear. The time delay estimation approach proposed here, is based on the use of local phase differences between specific frequency components of the received signal. Using this approach overcomes the need to cross-correlate the received signal with either a reference signal or the transmitted signal.The developed novel approach for time delay estimation, hence for distance and speed of sound measurement outperform the conventional correlation-based techniques and overcomes the 2π-phase ambiguity in the phase-based approaches and most practical situations can be accommodated (Assous et al., 2008; 2010).

3. Distance measurement using the received signal phase differences between components: new concept

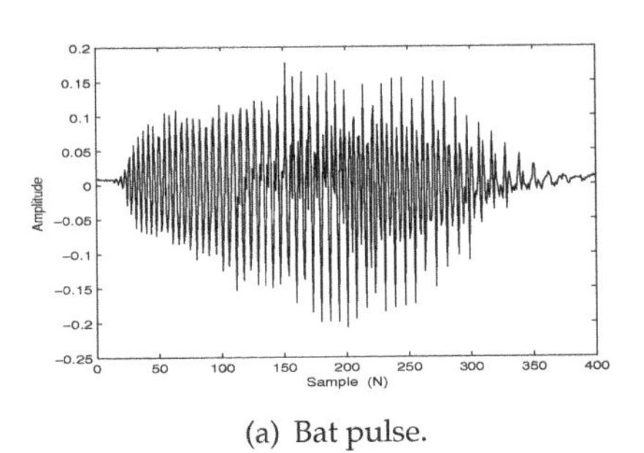

(a) Bat pulse.

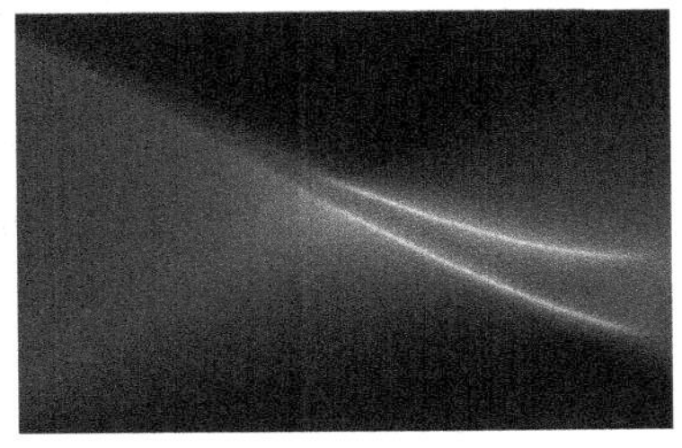

(b) Time-Frequency plot of the bat pulse.

Fig. 1

Consider the time-frequency plot of a single bat pulse shown in Fig.1, we note that at any particular time within the pulse there are essentially two frequencies present. The pulse length is about 2 ms and the frequency bandwidth is about 40 kHz. Let describe in the following how using two or more frequencies we may infer a distance.

To explain the concept, consider a scenario where an acoustic pulse contains a single frequency component f_1 with an initial zero phase offset. This pulse is emitted through the medium, impinges on a target, is reflected and returns back. The signal is captured and its phase measured relative to the transmitted pulse. Given this situation, we cannot estimate the distance to and from an object greater than one wavelength away (hence, usually, we would estimate the time of arrival of the pulse and assume a value for the velocity of sound in the medium to estimate the distance to the target).

For simplicity, assume the pulse contains a single cycle of frequency f_1 of wavelength λ_1.

The distance D to the target can be expressed as

$$D = n_1\lambda_1 + r_1 \tag{1}$$

where $\lambda_1 = v/f_1$, n_1 is an integer, r_1 is a fraction of the wavelength λ_1 and v is the speed of sound in the medium.

r_1 can be expressed as follows

$$r_1 = \lambda_1 \times \frac{\phi_1}{360} \tag{2}$$

where ϕ_1 is the residual phase angle in degrees. Combining equations (1) and (2) and rearranging

$$\begin{aligned} D &= n_1\lambda_1 + \lambda_1\frac{\phi_1}{360} \\ &= n_1\frac{v}{f_1} + \frac{\phi_1}{360}\frac{v}{f_1} \\ D &= \frac{v}{f_1}(n_1 + \frac{\phi_1}{360}) \end{aligned} \tag{3}$$

If we transmit a second frequency component f_2 within the same pulse, then it will also have associated with it a wavelength λ_2 and a residual phase ϕ_2, similarly:

$$D = \frac{v}{f_2}(n_2 + \frac{\phi_2}{360}) \tag{4}$$

Equations (1) and (2) can be solved by finding (2) $-$ (1) $\times$ $(\frac{\lambda_2}{\lambda_1})$ and rearranged to give

$$\begin{aligned} D &= (\frac{\lambda_1\lambda_2}{\lambda_1 - \lambda_2})((n_2 - n_1) + \frac{(\phi_2 - \phi_1)}{360}) \\ D &= (\frac{\lambda_1\lambda_2}{\lambda_1 - \lambda_2})(\Delta n + \frac{\Delta\phi}{360}) \end{aligned} \tag{5}$$

Using $v = f \times \lambda$ we obtain

$$D = \frac{v}{f_2 - f_1}((n_2 - n_1) + \frac{(\phi_2 - \phi_1)}{360}) = \frac{v}{\Delta f}(\Delta n + \frac{\Delta\phi}{360}) \tag{6}$$

Knowing $D = v \times t$ we deduce the time delay t as

$$t = \frac{1}{\Delta f}(\Delta n + \frac{\Delta\phi}{360}) \tag{7}$$

If we impose the condition that $\Delta n \leq 1$ then (6) can be solved. This restriction on Δn is imposed as follows:

- A distance D is chosen within which we require an unambiguous range measurement.
- Select a frequency f_1 within the bandwidth of the system, and its corresponding wavelength λ_1, $n_1 = \frac{D}{\lambda_1}$ (from (1)).
- Similarly, using (1), select frequency f_2 with its corresponding wavelength λ_2 such that the number of cycles is $n_2 = n_1 + 1$.

Considering (6), the maximum range is achieved by this approach when Δn=1; is

$$R = \frac{v}{\Delta f} \tag{8}$$

Therefore, R is the maximum unambiguous range that can be achieved using two frequencies f_1 and f_2 as described above, where any distance within the range R can be determined unambiguously.

3.1 Example

- Defining an unambiguous range $R = 1500$ mm and assuming the speed of sound in water v = 1500 m/s=1.5 mm/μs, selecting $f_1 = 200.0$ kHz, gives λ_1= 7.5 mm.
- Using (1), for this range $R = D$, n_1= 200.0 cycles, $r_1 = 0$. Ensuring Δn=1, from (6) requires n_2=201.0 cycles, and $f_2 = 201.0$ kHz.
- Consider a distance to target d= 1000.1234 mm, we wish to estimate (which is unknown but is within the unambiguous range R).
- Using frequencies f_1 and f_2 defined above, and equations (1) and (6), gives
  - $n_1 = 133$, r_1=0.349786 cycle, $\Longleftrightarrow \phi_1 = 0.349786 \times 360 = 125.923^\circ$.
- Similarly for the frequency f_2 we find the residual phase
  - $n_2 = 134$, r_2=0.0165356 $\Longleftrightarrow \phi_2 = 5.953^\circ$
  - Thus, $\Delta\phi$=$\phi_2 - \phi_1$ = 5.953-125.923 =-119.970°.

  We use this value in the formula given in (6). However, since $\Delta\phi$ is negative (this means Δn = 1), we add 360° giving 240.0296° (If $\Delta\phi$ was positive we would have used the value directly).
- Now, v = 1.500 mm/μs, $\Delta f = 1$ kHz, substituting into (7) gives a first estimate of the range $\hat{d}_{f_1 f_2} = 1000.1233$ mm.

  The Unambiguous Range R (8) is independent of the frequencies used, depending only on the difference in frequency Δf.
- Note that in practice such resolution may not be achievable and limitations must be considered. For example, if the uncertainty of estimating the phase is within $\pm 0.5^\circ$, then the phases in the example above become ϕ_1=126.0 and ϕ_2=6.0, giving d=1000.0 mm implying an error of 0.1234 mm.

3.2 Using multiple frequencies through a "Vernier approach"

In (6), we imposed the condition that $\Delta n \leq 1$. The values of frequencies f_1 and f_2 were chosen to insure this condition and to obtain a first estimate of the distance $\hat{d}_{f_1 f_2}$ (6), and an estimate of the time delay $\hat{t}_{f_1 f_2}$ (7).

Introducing a third frequency $f_3 = 210$ kHz, such that ($f_3 - f_1 = 10 \times (f_2 - f_1)$); f_2 differs from f_1 by 1 kHz and f_3 differs from f_1 by 10 kHz.

Again from (1), for f_3 and d=1000.1234 mm, $n_3 = 140$ cycles and $r_3 = 0.017276$. Thus, $\phi_3 = 6.219°$ which we would measure as 6.5°. Thus, $\Delta\phi_{13} = \phi_3 - \phi_1 = 6.5 - 126 = -119.5°$. We add 360° to give 240.5°. However, Δn_{13} between frequencies f_1 and f_3 is now 7 (in fact 6, since we have already added in 360° to make the phase difference positive).

Using (6) with $\Delta\phi_{13}$ and different values of Δn_{13} (0 to 6) to get different distance estimate $\hat{d}_{f_1 f_3}$.

Applying (6) recursively for $\Delta n_{13} = 0...6$ to calculate $\hat{d}^k_{f_1 f_3} |_{k=0,6}$ selecting $\hat{d}^k_{f_1 f_3}$ closest in value to $\hat{d}_{f_1 f_2}$ as the optimum value $\hat{d}_{f_1 f_3} = 1000.2083mm |_{k=6}$. Hence, a new best time delay estimate $\hat{t}_{f_1 f_3 |_{k=6}}$.

Note that the new best estimate distance is with an error of 0.0849 mm. If a 4th frequency is introduced $f_4 = 300.0$ kHz, such that the $\Delta f = f_4 - f_1 = 100.0$ kHz, using (1) again, gives n_4 = 200 cycles and $r_4 = 0.02468$ corresponding to $\phi_4 = 8.8848°$ which we measure as 9°. Thus, $\Delta\phi = 9 - 126 = -117°$ which gives $\Delta\phi = 249.5$ after adding 360°. Note that $\Delta n = 66$ in this case.

Similarly, select the estimate $\hat{d}_{f_1 f_4}$ (Δn =0,1,2,..66) close in value to $\hat{d}_{f_1 f_3}$. This occurs at $\Delta n = 66$ giving $\hat{d}_{f_1 f_4} = 1000.125$ mm. Taking this as the best estimate, the final error is 0.0016 mm = 1.6 microns.

Thus, this example is reminiscent of the operation of a Vernier gauge as follows:

- $\Delta\phi_{12}$, related to the frequencies f_1 and f_2, gives the first estimate of the distance $\hat{d}_{f_1 f_2}$, hence $\hat{t}_{f_1 f_2}$.
- A higher frequency f_3 is then used (decade difference) to measure the same range but with a finer resolution. So a more accurate approximation to the measured range is obtained $\hat{d}_{f_1 f_3}$.
- Similarly, the measured range $\hat{d}_{f_1 f_4}$ corresponding to $\Delta\phi_{14}$ within f_1 and f_4, will give the ultimate estimate of the measured range d.
- Consequently, the maximum distance and the minimum resolution achieved are determined by the choice of the frequencies f_1, f_2, f_3 and f_4.

3.3 Phase offset measurement calibration

In the numerical example above, it is assumed the phases are faithfully transmitted and received, with no phase error on transmission or reception. It is also assumed that all frequencies have zero phase offset with respect to each other. In practice this is almost certainly not the case and such phase offsets between frequencies should be accounted for as discussed below.

Considering two frequencies $f_1 = 200.0$ kHz and $f_2 = 201$ kHz, assuming the speed of sound in water (v =1.5 mm/μs), from (8), the unambiguous range $R = 1500$ mm and Δn is 0 or 1. Considering the above

$$t = \frac{D}{v} = \frac{n + \phi/360}{f} \tag{9}$$

Consider two distances d_1, d_2 corresponding to two "times" t_1 and t_2 such that the number of cycles n is the same for both frequencies over these distances, and assume the phase measured includes a phase offset for that frequency. As an example, suppose the unknown phase offset for f_1 is 10°, for f_2 is 30° and assume d_1 = 100 mm.

From (9), the term $(n_1 + \phi_1/360)$ would be calculated as 13.3333 cycles, where $\phi_1 = 120°$. The 'measured' $\phi_1 = 120 + 10 = 130°$ $(\phi_{1measured} = \phi_{1distance} + \phi_{1offset})$.

Similarly, for f_2 we obtain $(n_2 + \phi_2/360) = 13.40$ cycles, where $\phi_2 = 144°$. The 'measured' $\phi_2 = 144 + 30 = 174°$; from (7), $t_1 = \Delta\phi/(360 \times \Delta f) = \frac{(174-130)}{360 \times \Delta f}$ =12.2222 μs. The actual time should be 6.6666 μs.

Assume a second distance d_2= 200 mm. Using (9), for f_1 we obtain $(n_1 + \phi_1/360) = 26.6666$ cycles, which gives $\phi_1 = 240°$, the 'measured' $\phi_1 = 240 + 10 = 250°$. For f_2 we obtain $(n_2 + \phi_2/360)$ =26.80 cycles, which gives $\phi_2 = 288°$. The 'measured' $\phi_2 = 288 + 30 = 318°$.

Thus, using (7), $t_2 = \Delta\phi/(360 \times \Delta f) = \frac{(318-250)}{360\Delta f} = 18.8888$ μs. The actual time should be 13.3333 μs.

Plotting d as the x-axis and t as the y-axis we obtain a linear relationship

$$t = 0.6666 \times d + 5.5555 \tag{10}$$

where the slope (0.6666=1/1.5) is the speed of sound measured as 1 mm per 0.6666 μs or 1/0.6666 = 1.5 mm/μs. The intercept (5.5555 μs) is a measure of the relative phase between f_1 and f_2. Since $\Delta f = 10$ kHz, 1 cycle is 100 μs long, consequently the offset of 5.5555 $\mu s \equiv 360 \times (5.5555/100) = 20°$ which is equal to the relative phase (30-10) between the two frequencies. If we had known the phase offset between the two frequencies (20°) then in the calculation of times we would have obtained for t_1 a new phase difference of $(174 - 130 - 20) = 24°$ giving a time for $t_1 = 6.6666$ μs.
Similarly, for t_2 we obtain a new phase difference of $(318 - 250 - 20) = 48°$ giving a time for $t_2 = 13.3333$ μs. Both t_1 and t_2 are now correct.

Note that:

- If we assumed d_1 was 100 mm but it was actually say 120 mm and that d_2 was 200 mm but it was actually 220 mm, then we obtain the phase offset as 15.2°, the slope of (10) above however, is unaffected. For example, such uncertainty may arise if the distance traveled by the wave within the transducers is not taken into consideration.
- If the temperature changes and so v changes, this changes the slope of (10) but not the time intercept or the phase offset. For example, if v=1.6 mm/μs, then equation (10) becomes $t = (1/1.6) \times d + 5.5555 = 0.625 \times d + 5.5555$.

4. Application

4.1 Experiment

To demonstrate this approach, a series of measurements were performed in a water tank measuring 1530 × 1380 × 1000 mm^3. Two broadband ultrasonic transducers were used, having a wide bandwidth with a centre frequency between 100 kHz and 130 kHz. They operate as both transmitters (Tx) and receivers (Rx) of ultrasound with a beam width of around 10 degrees at the centre frequency, where -3dB bandwidth is 99 kHz (72 kHz to 171 kHz). The transducers were mounted on a trolley, moveable in the Y-direction, which was in-turn mounted on a rail, moveable in the X-direction. The experimental set-up is illustrated in Fig. 2.

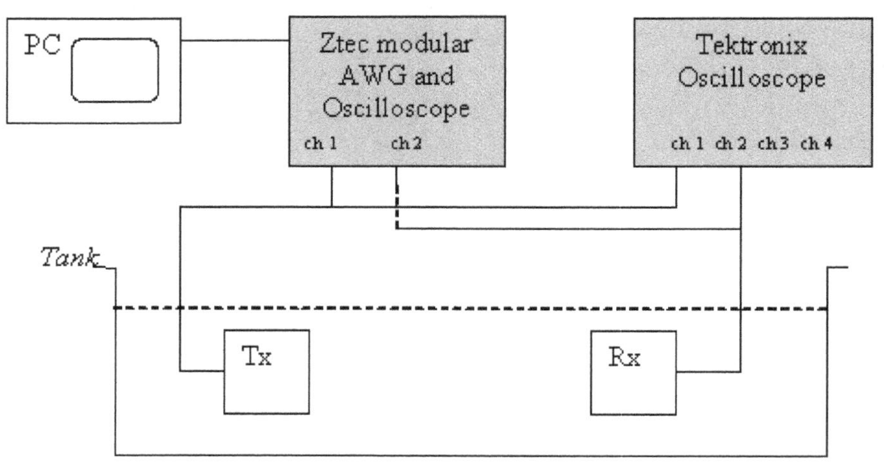

Fig. 2. Schematic diagram of the experimental setup

For this purpose, linear encoders were used to measure displacement of the rails in the x-direction. Software written in Visual Basic provided readouts of transducer positions. The temperature in the tank was measured by thermocouples which were calibrated using a quartz thermometers which are traceable to national standard. They were positioned on the four sides panels of the tank and recorded 19.84, 19.89, 19.89 and 19.88°C during the experiment.

The transmitter was driven directly by a 20 V peak-to-peak waveform consisting of four sine waves with zero phase offset (70 kHz, 71 kHz, 80 kHz and 170 kHz) added together. A modular system comprising a 16-bit arbitrary waveform generator (Ztec ZT530PXI) and a 16-bit digital storage oscilloscope (Ztec ZT410PXI) were used to transmit and receive signals. A program written in C++ was used to control the signal transmission and acquisition. Distances between Tx and Rx of 612.859 mm, 642.865 mm, 702.876 mm, 762.875 mm, 822.847 mm , 882.875 mm, 942.863 and 999.804 mm were chosen to be within the unambiguous range $R \approx 1500$ mm (8), as set by the linear encoders. A set of 10 signals; were transmitted for each distance described above. Note that, before transmitting, each signal was multiplied by a Tukey window (cosine-tapered window with a 0.1 taper ratio) to reduce the "turn on" and "turn off" transients of the transducers.

At each distance, 3 repetitive pulses were transmitted and received for each distance. Furthermore, 6 repetitive pulses were transmitted and received while keeping the distance constant at 999.804 mm to assess the repeatability of the system (see Fig.3). Each pulse was 3 ms long containing the 4 added frequency components described above. The sampling frequency Fs was set to 10 MHz, giving a number of samples N=20000. A Discrete Fourier Transform (DFT) was then applied to the received pulses to obtain the magnitude and phase information for each of the 4 frequency components, using a window of $[\frac{N}{2}+1:N]$ for each received signal. This gave a resolution $\frac{Fs}{N/2}$=1 kHz which was consistent with the smallest step between the 4 frequencies.

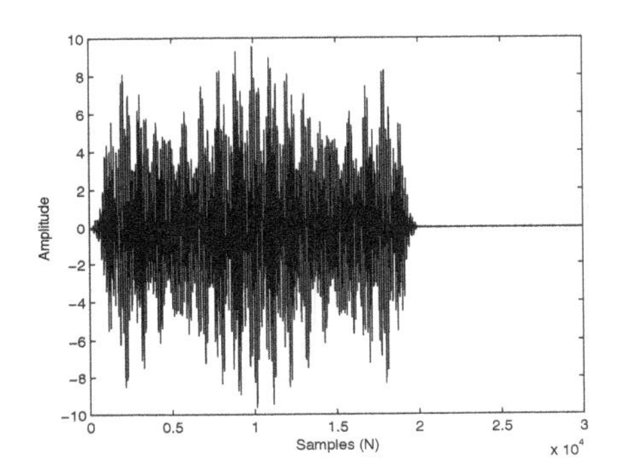

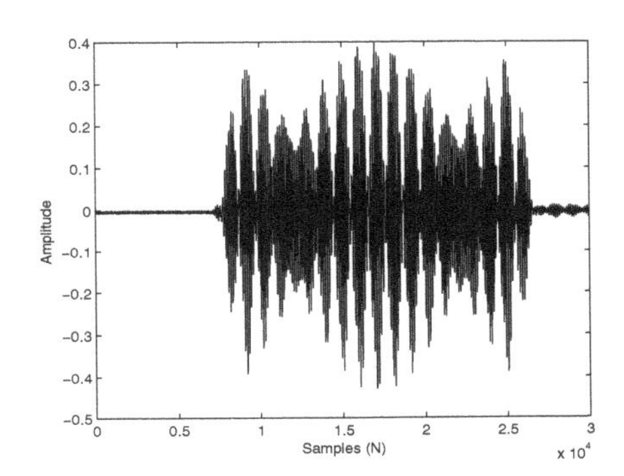

Fig. 3. Examples of transmitted and received signals.

Fig. 3 shows the original transmitted (left) and received (right) signals when Tx and Rx were 999.804 mm apart.

Note the DFT reports phase with respect to cosine, whereas sine waves were used in this experiment. Sine waves are returned with a phase of $-90°$ relative to cosine waves by the DFT. This was not an issue, since relative phase differences were used.

Distance (mm)	Estimated time (μs)
612.85900	446.557
642.86500	466.779
702.87600	507.267
762.87500	547.739
822.84700	587.975
882.87500	628.386
942.86300	668.855
999.80400	707.183

Table 1. The estimated time delays for the eight distances

4.2 Results and discussion

Using the phase difference for each distance, the phase-based time delay approach was applied to obtain the corresponding estimated times for each phase difference $\Delta\phi_{12}$, $\Delta\phi_{13}$,

$\Delta\phi_{14}$, $\Delta\phi_{23}$, $\Delta\phi_{24}$ and $\Delta\phi_{34}$, for the pairs f_1f_2, f_1f_3, f_1f_4, f_2f_3, f_2f_4 and f_3f_4, respectively. Note that a careful use of the Fourier transform (DFT) have to be considered to calculate the phase offset for each component. To do this, we have to take into account the fact that all the frequency components have to be integer number of cycles and have to be in the frequency bins defined by the smallest frequency difference between components. Using a simple calculation of the first estimate by $t_{12} = \Delta\phi_{12}/(f_2 - f_1)$ as a first estimate using (7), gave corresponding estimated times $\hat{t}_{12}$, $\hat{t}_{13}$, $\hat{t}_{14}$, $\hat{t}_{23}$, $\hat{t}_{24}$ and $\hat{t}_{34}$, respectively. For each distance, $\hat{t}_{14}$ should be the best estimate (i.e. the greatest Δf). Note that, $\hat{t}_{14}$ parameter is taken as a best estimate for time delay measurement in Table.1. Fig. 3 shows the estimated $\hat{t}_{14}$ for the distance

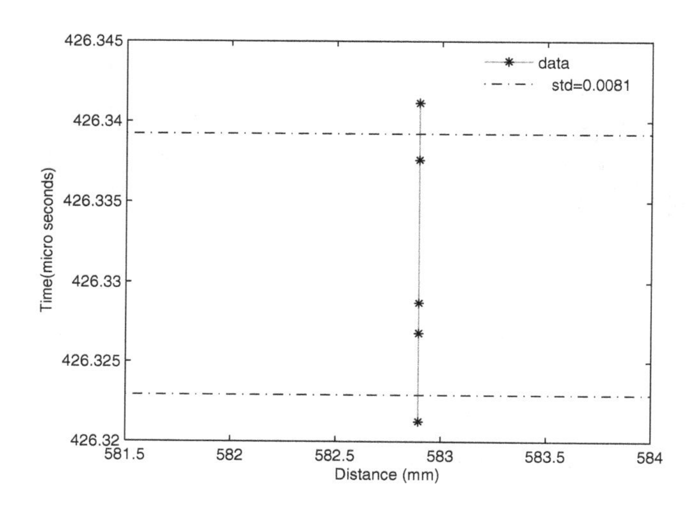

Fig. 4. Repeated time delay estimations for a transducer separation of 582.893 mm showing a maximum variability of almost 8 ns in the time delays estimated (617.8±0.0081 μs).

between Tx and Rx of 582.893 for 5 repeat pulses. The repeatability is shown to be within 8 ns.

5. Conclusion

In this chapter, a high resolution time delay estimation approach based on phase differences between components of the received signal has been proposed. Hence no need to do a cross-correlation between the transmitted and the received signal as all the information is comprised in the received signal in the local phase between components. Within an unambiguous range defined by the smallest and the highest frequency in the signal, any distance can be estimated. This time delay technique leads also to high resolution distance and speed of sound measurement. This approach is tolerant to additive gaussian noise as it relies on local phase difference information. The technique is costly effective as it relies on software and no need for local oscillator to overcome such phase ambiguity by counting the integer number of cycles in the received signal as it is done by the conventional methods. We have mentioned that, authors in Whitlow & Simmons (2007), concluded that, bat can achieve a resolution of $20\mu m$ in air, if we can measure the phase to an accuracy of 1 degree then this would allow us to get a resolution of 4mm/360=11μm, using the same wavelength as the bat do.

6. Acknowledgment

This work was undertaken as part of the Biologically Inspired Acoustic Systems (BIAS) project that is funded by the RCUK via the Basic Technology Programme grant reference number EP/C523776/1. The BIAS project involves collaboration between the British Geological Survey, Leicester University, Fortkey Ltd., Southampton University, Leeds University, Edinburgh University and Strathclyde University.

7. References

Thomas, J.A. & Moss, C.F. (2004). Echolocation in bats and dolphins, *University of Chicago Press.*

Whitlow, W.L. (1993). The Sonar of Dolphins, *Springer.*

Whitlow, W.L; Simmons. J. A (2007). Echolocation in dolphins and bats, *Physics Today*, page numbers (40-45).

Assous, S.; Jackson, P.; Hopper, C.; Gunn, D.; Rees, J.; Lovell, M. (2008). Bat-inspired distance measurement using phase information. *J Acoust Soc Am*, Vol. 124, page numbers (2597).

Assous, S.; Hopper, C.; Gunn, D.; Jackson, P.; Rees, J.; Lovell, M. (2010). Short pulse multi-frequency phase-based time delay estimation, *J. Acoust. Soc. Am.* Vol. 01, page numbers (309-315).

Knapp, G.H. & Carter, G.C. (1976). The generalised correlation method for estimation of time delay, *IEEE Trans. Acoust., Speech, Signal Processing*, Vol. ASSP-24, No.4, page numbers (320-327).

Carter, G.C. (1979). Sonar signal processing for source state estimation. *IEEE Pub. 79CH1476-1-AES*, page numbers (386-395).

Carter, G.C. (1987). Coherence and time delay estimation. *Proc. of the IEEE*, Vol. 75, No.2, page numbers (236-255).

Boucher, R.E. & Hassab, J.C (1981). Analysis of discrete implementation of generalised cross-correlator. *IEEE Trans. Acoust., Speech, Signal Processing*, Vol. ASSP-29, page numbers (609-611).

Chen, J.; Benesty, J.; Huang, Y.A.; (2004). Time delay estimation via linear interpolation and cross correlation, *IEEE Trans. Speech and audio Processing*, Vol. 12, No.5, page numbers (509-519).

Chen, J.; Benesty, J.; Huang, Y.A. (2006). Time delay estimation in room acoustic environments: An Overview. *Eurasip Journal on Applied Signal Processing*, Vol. 2006, Article ID 26503, page numbers (1-19).

Belostotski, L.; Landecker, T.L.; Routledge, D.(2001). Distance measurement with phase stable CW radio link unsing the Chinese remainder theorem. *Electronics Letters*, Vol.37, No.8, page numbers (521-522).

Towers, C.E.; Towers, P.D.; Jones-Julian, D.C. (2004). The efficient Chinese remainder theorem algorithm for full-field fringe phase analysis in multi-wavelength interferometry. *Optics Express*, Vol. 12, No.6, page numbers (1136-1143).

Xiang-Gen, X.; Kenjing, L. (2005). A generalized Chinese reminder theorem for residue sets with errors and its application in frequency determination from multiple sensors with low sampling rates, *IEEE Process. Lett.*, Vol. 12, page numbers (768-771).

Goldreich, O.; Ron, D.; Sudan, M. (2000). Chinese remindering with errors, *IEEE Trans. Inf. Theory*, Vol. 93, page numbers (1330-1338).

Lee, A.K.T.; Lucas, J.; Virr, L.E (1989). Microcomputer-controlled acoustic range finding technique. *J. Phys. E: Sci. Instrum.*, Vol.22, page numbers (52-58).

Yang, M.; Hill, S.L.; Bury, B.; Gray, J.O (1994). A multifrequency AM-based ultrasonics system for accuracy distance measurement. *IEEE Trans. Instrumentation and measurement*, Vol. 43, No.6, page numbers (861-866).

2

Localization of Buried Objects in Sediment Using High Resolution Array Processing Methods

Caroline Fossati, Salah Bourennane and Julien Marot
Institut Fresnel, Ecole Centrale Marseille
France

1. Introduction

Non-invasive range and bearing estimation of buried objects, in the underwater acoustic environment, has received considerable attention (Granara et al., 1998).
Many studies have been recently developed. Some of them use acoustic scattering to localize objects by analyzing acoustic resonance in the time-frequency domain, but these processes are usually limited to simple shaped objects (Nicq & Brussieux, 1998). In (Guillermin et al., 2000) the inversion of measured scattered acoustical waves is used to image buried object, but the frequencies used are high and the application in a real environment should be difficult. The acoustic imagery technique uses high frequencies that are too strongly attenuated inside the sediment therefore it is not suitable. Another method which uses a low frequency synthetic aperture sonar (SAS) has been applied on partially and shallowly buried cylinders in a sandy seabed (Hetet et al., 2004). Other techniques based on signal processing such as time reversal technic (Roux & Fink, 2000), have been also developed for object detection and localization but their applicability in real life has been proven only on cylinders oriented in certain ways and point scatterers. Furthermore, having techniques that operate well for simultaneous range and bearing estimation using wideband and fully correlated signals scattered from nearfield and farfield objects, in a noisy environment, remains a challenging problem.
In this chapter, the proposed method is based on array processing methods combined with an acoustic scattering model. Array processing techniques, as the MUSIC method, have been widely used for acoustic point sources localization. Typically these techniques assume that the point sources are on the seabed and are in the farfield of the array so that the measured wavefronts are all planar. The goal then is to determine the direction of arrival (bearing) of these wavefronts. These techniques have not been used for bearing and range estimation of buried objects and in this chapter we are interested to extend them to this problem. This extension is a challenging problem because here the objects are not point sources, are buried in the seabed and can be everywhere (in the farfield or in the nearfield array). Thus the knowledge of the bearing is not sufficient to localize the buried object. Furthermore, the signals are correlated and the Gaussian noise should be taken into account. In addition we consider that the objects have known shapes. The principal parameters that disturb the object localization problem, are the noise, the lack of knowledge of the scattering model and the presence of correlated signals. In the literature there is any method able to solve all those parameters. However we can found a satisfying method to cope with each parameter (noise,

correlated signals and lack of knowledge of the scattering model) alone, thus we have selected the following methods,

- High order statistics are famous by their power to remove the additive Gaussian noise and then to clean the data. It consists in using the slice cumulant matrix instead of using the spectral matrix and operate at a fixed frequency (narrowband signal). It has been employed in the MUSIC method in order to estimate the bearing sources (Gönen & Mendel, 1997), (Mendel, 1991),
- The frequential smoothing is a technique which allows us to decorrelate the wideband signals (Valaee & Kabal, 1995) by means of an average of the focused spectral matrices formed for all the frequencies of the frequency band. It has been employed also in the MUSIC method in order to estimate the bearing sources,
- The exact solution of the acoustic scattering model has been addressed in many published work for several configurations, as single (Doolittle & Uberall, 1966), (Goodman & Stern, 1962) or multiple objects (Prada & Fink, 1998), (Zhen, 2001), buried or partially buried objects (Lim et al., 1993), (Tesei et al., 2002), with cylindrical (Doolittle & Uberall, 1966), (Junger, 1952), or spherical shape (Fawcett et al., 1998), (Goodman & Stern, 1962), (Junger, 1952),

In this chapter we propose to adapt array processing methods and acoustic scattering model listed above in order to solve the problem of burried object with known shape by estimating their bearing and range, considering wideband correlated signals in presence of Gaussian noise. The fourth-order cumulant matrix (Gönen & Mendel, 1997), (Mendel, 1991) is used instead of the cross-spectral matrix to remove the additive Gaussian noise. The bilinear focusing operator is used to decorrelate the signals (Valaee & Kabal, 1995) and to estimate the coherent signal subspace (Valaee & Kabal, 1995), (Wang & Kaveh, 1985). From the exact solution of the acoustic scattered field (Fawcett et al., 1998), (Junger, 1952), we have derived a new source steering vector including both the range and the bearing objects. This source steering vector is employed in MUSIC (MUltiple SIgnal Classification) algorithm (Valaee & Kabal, 1995) instead of the classical plane wave model.

The organization of this chapter is as follows : the problem is formulated in Section 2. In Section 3, the scattering models are presented. In Section 4, the cumulant based coherent signal subspace method for bearing and range estimation is presented. Experimental setup and the obtained results supporting our conclusions and demonstrating our method are provided in Sections 5 and 6. Finally, conclusions are presented in Section 7.

Throughout the chapter, lowercase boldface letters represent vectors, uppercase boldface letters represent matrices, and lower and uppercase letters represent scalars. The symbol "T" is used for transpose operation, the superscript "+" is used to denote complex conjugate transpose, the subscript "*" is used to denote conjugate operation, and $||.||$ denotes the L_2 norm for complex vectors.

2. Problem formulation

We consider a linear array of N sensors which received the wideband signals scattered from P objects ($N > P$) in the presence of an additive Gaussian noise. The received signal vector, in the frequency domain, is given by (Gönen & Mendel, 1997), (Mendel, 1991)

$$\mathbf{r}(f_n) = \mathbf{A}(f_n)\mathbf{s}(f_n) + \mathbf{b}(f_n), \text{ for } n = 1, ..., L \tag{1}$$

where,

$$\mathbf{A}(f_n) = [\mathbf{a}(f_n, \theta_1, \rho_1), \mathbf{a}(f_n, \theta_2, \rho_2), ..., \mathbf{a}(f_n, \theta_P, \rho_P)], \tag{2}$$

$$\mathbf{s}(f_n) = [s_1(f_n), s_2(f_n), ..., s_P(f_n)]^T, \tag{3}$$

$$\mathbf{b}(f_n) = [b_1(f_n), b_2(f_n), ..., b_N(f_n)]^T. \tag{4}$$

$\mathbf{r}(f_n)$ is the Fourier transforms of the array output vector, $\mathbf{s}(f_n)$ is the vector of zero-mean complex random non-Gaussian source signals, assumed to be stationary over the observation interval, $\mathbf{b}(f_n)$ is the vector of zero-mean complex white Gaussian noise and statistically independent of signals and $\mathbf{A}(f_n)$ is the transfer matrix (steering matrix) of the source sensor array systems computed by the $\mathbf{a}(f_n, \theta_k, \rho_k)$ for $k = 1, ..., P$, object steering vectors, assumed to have full column rank, it is given by:

$$\mathbf{a}(f_n, \theta_k, \rho_k) = [a(f_n, \theta_{k1}, \rho_{k1}), a(f_n, \theta_{k2}, \rho_{k2}), ..., a(f_n, \theta_{kN}, \rho_{kN})]^T, \tag{5}$$

where θ_k and ρ_k are the bearing and the range of the kth object to the first sensor of the array, thus, $\theta_k = \theta_{k1}$ and $\rho_k = \rho_{k1}$. In addition to the model equation (1), we also assume that the signals are statistically independent. In this case, a fourth order cumulant is given by

$$\mathrm{Cum}(r_{k_1}, r_{k_2}, r_{l_1}, r_{l_2}) = \mathrm{E}\{r_{k_1} r_{k_2} r_{l_1}^* r_{l_2}^*\} - \mathrm{E}\{r_{k_1} r_{l_1}^*\}\mathrm{E}\{r_{k_2} r_{l_2}^*\} - \mathrm{E}\{r_{k_1} r_{l_2}^*\}\mathrm{E}\{r_{k_2} r_{l_1}^*\}$$

where r_{k_1} is the k_1 element in the vector $\mathbf{r}$ and where E[.] denotes the expectation operator. The indices k_1, k_2, l_1, l_2 are similarly defined. The cumulant matrix consisting of all possible permutations of the four indices $\{k_1, k_2, l_1, l_2\}$ is given in (Yuen & Friedlander, 1997) as

$$\mathbf{C}(f_n) = \sum_{k=1}^{P} \Big(\mathbf{a}(f_n, \theta_k, \rho_k) \otimes \mathbf{a}^*(f_n, \theta_k, \rho_k)\Big) u_k(f_n) \Big(\mathbf{a}(f_n, \theta_k, \rho_k) \otimes \mathbf{a}^*(f_n, \theta_k, \rho_k)\Big)^+ \tag{6}$$

where $u_k(f_n)$ is the source kurtosis (i.e., fourth order analog of variance) defined by

$$u_k(f_n) = \mathrm{Cum}(s_k(f_n), s_k^*(f_n), s_k(f_n), s_k^*(f_n)) \tag{7}$$

of the kth complex amplitude source and $\otimes$ is the Kronecker product. When there are N array sensors, $\mathbf{C}(f_n)$ is $(N^2 \times N^2)$ matrix. The rows of $\mathbf{C}(f_n)$ are indexed by $(k_1 - 1)N + l_1$, and the columns are indexed by $(l_2 - 1)N + k_2$. In terms of the vector $\mathbf{r}(f_n)$, the cumulant matrix $\mathbf{C}(f_n)$ is organized compatibly with the matrix, $\mathrm{E}\{(\mathbf{r}(f_n) \otimes \mathbf{r}^*(f_n))(\mathbf{r}(f_n) \otimes \mathbf{r}^*(f_n))^+\}$. In other words, the elements of $\mathbf{C}(f_n)$ are given by:

$$\mathbf{C}((k_1 - 1)N + l_1, (l_2 - 1)N + k_2) \tag{8}$$

for $k_1, k_2, l_1, l_2 = 1, 2, ..., N$ and

$$\mathbf{C}((k_1-1)N+l_1,(l_2-1)N+k_2) = \mathrm{Cum}(r_{k_1}, r_{k_2}, r_{l_1}, r_{l_2}) \tag{9}$$

where r_i is the ith element of the vector $\mathbf{r}$. In order to reduce the calculating time, instead of using the cumulant matrix $\mathbf{C}(f_n)$, a cumulant slice matrix $(N \times N)$ of the observation vector at frequency f_n can be calculated and it offers the same algebraic properties of $\mathbf{C}(f_n)$. This matrix is denoted $\mathbf{C}_1(f_n)$ (Gönen & Mendel, 1997), (Yuen & Friedlander, 1997). If we consider a cumulant slice, for example, by using the first row of $\mathbf{C}(f_n)$ and reshape it into an $(N \times N)$ hermitian matrix (Bourennane & Bendjama, 2002), i.e.

$$\begin{aligned}\mathbf{C}_1(f_n) =& \mathrm{Cum}(r_1(f_n), r_1^*(f_n), \mathbf{r}(f_n), \mathbf{r}^+(f_n)) \\ =& \begin{bmatrix} c_{1,1} & c_{1,N+1} & \cdots & c_{1,N^2-N+1} \\ c_{1,2} & c_{1,N+2} & \cdots & c_{1,N^2-N+2} \\ \vdots & \vdots & \vdots & \vdots \\ c_{1,N} & c_{1,2N} & \cdots & c_{1,N^2} \end{bmatrix} \\ =& \mathbf{A}(f_n)\mathbf{U}_s(f_n)\mathbf{A}^+(f_n)\end{aligned} \tag{10}$$

where $c_{1,j}$ is the $(1,j)$ element of the cumulant matrix $\mathbf{C}(f_n)$ and $\mathbf{U}_s(f_n)$ is the diagonal kurtosis matrix, its ith element is defined as, $\mathrm{Cum}(s_i(f_n), s_i^*(f_n), s_i(f_n), s_i^*(f_n))$ with $i = 1, ..., P$.
$\mathbf{C}_1(f_n)$ can be reported as the classical covariance or spectral matrix of received data

$$\mathbf{\Gamma}_r(f_n) = \mathrm{E}\left[\mathbf{r}(f_n)\mathbf{r}^+(f_n)\right] = \mathbf{A}(f_n)\mathbf{\Gamma}_s(f_n)\mathbf{A}^+(f_n) + \mathbf{\Gamma}_b(f_n) \tag{11}$$

where $\mathbf{\Gamma}_b(f_n) = \mathrm{E}\left[\mathbf{b}(f_n)\mathbf{b}^+(f_n)\right]$ is the spectral matrix of the noise vector and

$$\mathbf{\Gamma}_s(f_n) = \mathrm{E}\left[\mathbf{s}(f_n)\mathbf{s}^+(f_n)\right] \tag{12}$$

is the spectral matrix of the complex amplitudes $\mathbf{s}(f_n)$.
If the noise is white then:

$$\mathbf{\Gamma}_b(f_n) = \sigma_b^2(f_n)\mathbf{I}, \tag{13}$$

where $\sigma_b^2(f_n)$ is the noise power and $\mathbf{I}$ is the $(N \times N)$ identity matrix. The signal subspace is shown to be spanned by the P eigenvectors corresponding to P largest eigenvalues of the data spectral matrix $\mathbf{\Gamma}_r(f_n)$. But in practice, the noise is not often white or its spatial structure is unknown, hence the interest of the high order statistics as shown in equation (4) in which the fourth order cumulants are not affected by additive Gaussian noise (i.e., $\mathbf{\Gamma}_b(f_n) = 0$), so as no noise spatial structure assumption is necessary. If the eigenvalues and the corresponding eigenvectors of $\mathbf{C}_1(f_n)$ are denoted by $\{\lambda_i(f_n)\}_{i=1..N}$ and $\{\mathbf{v}_i(f_n)\}_{i=1..N}$. Then, the eigendecomposition of the cumulant matrix $\mathbf{C}_1(f_n)$ is exploited so as

$$\mathbf{C}_1(f_n) = \sum_{i=1}^{N} \lambda_i(f_n)\mathbf{v}_i(f_n)\mathbf{v}_i^+(f_n) \tag{14}$$

In matrix representation, equation (14) can be written

$$\mathbf{C}_1(f_n) = \mathbf{V}(f_n)\Lambda(f_n)\mathbf{V}^+(f_n) \tag{15}$$

where

$$\mathbf{V}(f_n) = [\mathbf{v}_1(f_n), ..., \mathbf{v}_N(f_n)] \tag{16}$$

and

$$\mathbf{\Lambda}(f_n) = diag(\lambda_1(f_n), ..., \lambda_N(f_n)). \tag{17}$$

Assuming that the columns of $\mathbf{A}(f_n)$ are all different and linearly independent it follows that for nonsingular $\mathbf{C}_1(f_n)$, the rank of $\mathbf{A}(f_n)\mathbf{U}_s(f_n)\mathbf{A}^+(f_n)$ is P. This rank property implies that:

- the $(N-P)$ multiplicity of its smallest eigenvalues : $\lambda_{P+1}(f_n) = \ldots = \lambda_N(f_n) \cong 0$.
- the eigenvectors $\mathbf{V}_b(f_n) = \{\mathbf{v}_{P+1}(f_n)\ldots\mathbf{v}_N(f_n)\}$ corresponding to the minimal eigenvalues are orthogonal to the columns of the matrix $\mathbf{A}(f_n)$, namely, the steering vectors of the signals

$\mathbf{V}_b(f_n) = \{\mathbf{v}_{P+1}(f_n)\ldots\mathbf{v}_N(f_n)\}\perp\{\mathbf{a}(f_n,\theta_1,\rho_1)\ldots\mathbf{a}(f_n,\theta_P,\rho_P)\}$
The eigenstructure based techniques are based on the exploitation of these properties. The spatial spectrum of the conventional MUSIC method can be modified as follows in order to estimate both the range and the bearing of objects at the frequency f_n,

$$Z(f_n,\theta_k,\rho_k) = \frac{1}{\mathbf{a}^+(f_n,\theta_k,\rho_k)\mathbf{V}_b(f_n)\mathbf{V}_b^+(f_n)\mathbf{a}(f_n,\theta_k,\rho_k)} \tag{18}$$

Then the location (θ_k,ρ_k) for $k = 1,...,P$, maximizing the modified MUSIC spectrum in (18) is selected as the estimated object center to the first sensor of the array. Because a two dimensional search requires that the exact solution of the scattered field be calculated at each point in order to fill the steering vector $\mathbf{a}(f_n,\theta_k,\rho_k)$.

3. The scattering model

In this section we present how to fill the steering vector used in equation (18) at a fixed frequency f_n. Consider the case in which a plane wave is incident, with an angle θ_{inc}, on P infinite elastic cylindrical shells or elastic spherical shells of inner radius β_k and outer radius α_k for $k = 1,...,P$, located in a free space at (θ_k,ρ_k) the bearing and the range of the kth object, associated to the first sensor of the array $\mathbf{S}_1$ (see figure 1). The fluid outside the shells is labeled by 1, thus, the sound velocity c_1 and the wavenumber

$$K_{n1} = \frac{2\pi f_n}{c_1}.$$

3.1 Cylindrical shell

We consider the case of infinitely long cylindrical shell. In order to calculate the exact solution for the acoustic scattered field $a_{cyl}(f_n,\theta_{k1},\rho_{k1})$ a decomposition of the different fields is used, according to the Bessel functions J_m, N_m and the Hankel function (H_m).

The scattered pressure in this case is given by (Doolittle & Uberall, 1966),(Junger, 1952),

$$a_{cyl}(f_n, \theta_{k1}, \rho_{k1}) = p_{c0} \sum_{m=0}^{\infty} j^m \epsilon_m b_m H_m^{(1)}(K_{n1}\rho_{k1}) \cos(m(\theta_{k1} - \theta_{inc})), \quad (19)$$

where p_{c0} is a constant, $\epsilon_0 = 1, \epsilon_1 = \epsilon_2 = ... = 2$, b_m is a coefficient depending on conditions limits and m is the number of modes (Doolittle & Uberall, 1966).

3.2 Spherical shell

The analysis is now extended to the case where the scatterer is a spherical shell. The scattered pressure is given by (Fawcett et al., 1998), (Goodman & Stern, 1962), (Junger, 1952)

$$a_{sph}(f_n, \theta_{k1}, \rho_{k1}) = p_{s0} \sum_{m=0}^{\infty} B_m H_m^{(1)}(K_{n1}\rho_{k1}) P_m(\cos(\theta_{k1} - \theta_{inc})), \quad (20)$$

where p_{s0} is a constant and $P_m(\cos(\theta_{k1} - \theta_{inc}))$ is the Legendre polynomials (Goodman & Stern, 1962). Equations (19) and (20) give the first component of the steering vector, then, in a similar manner the other component $a_{cyl}(f_n, \theta_{ki}, \rho_{ki})$ and $a_{sph}(f_n, \theta_{ki}, \rho_{ki})$ for $i = 2, ..., N$, associated to the *ith* sensor, can be formed, where all the couples (θ_{ki}, ρ_{ki}) are calculated using the general Pythagore theorem and are function of the couple (θ_{k1}, ρ_{k1}). Thus, the configuration used is shown in figure 1. The obtained θ_{ki}, ρ_{ki} are given by

$$\rho_{ki} = \sqrt{\rho_{ki-1}^2 - d^2 - 2\rho_{ki-1} d \cos(\frac{\pi}{2} + \theta_{ki-1})} \quad (21)$$

$$\theta_{ki} = \cos^{-1}[\frac{d^2 + \rho_{ki}^2 - \rho_{ki-1}^2}{2\rho_{ki-1} d}], \quad (22)$$

where d is the distance between two adjacent sensors. Finally the steering vector is filled with the cylindrical scattering model in the case of cylindrical shells and filled with the spherical scattering model in the case of spherical shells. For example, when the considered objects are cylindrical shells, the steering vector is written as:

$$\mathbf{a}(f_n, \theta_k, \rho_k) = \left[a_{cyl}(f_n, \theta_{k1}, \rho_{k1}), \; ..., \; a_{cyl}(f_n, \theta_{kN}, \rho_{kN})\right]^T, \quad (23)$$

By using the exact solution of the scattered field (Doolittle & Uberall, 1966), we can fill the direction vector in the MUSIC algorithm equation (18) with non planar scattered field to locate the objects. The location $(\hat{\rho}, \hat{\theta})$ maximizing the MUSIC spectrum in (18) is selected as the estimated object center.
Because a two dimensional search requires that the exact scattered field be calculated at each point. The modified MUSIC algorithm presented in this section, is limited to one or multiple objects localization where the interactions are ignored. So, the localization problem is approached as if these objects are independently scattering the incident plane wave.

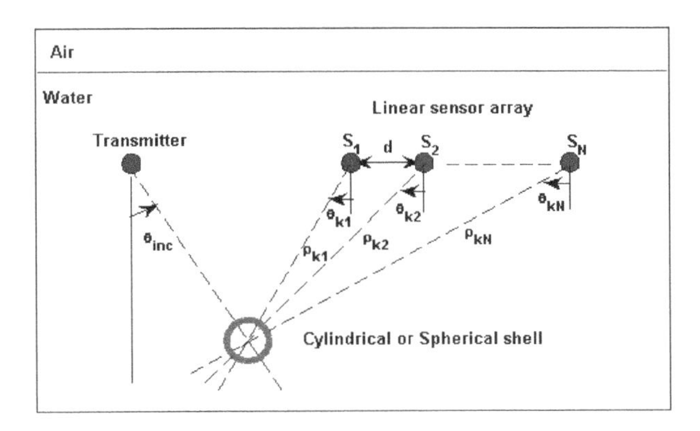

Fig. 1. Geometry configuration of the *kth* object localization.

4. The cumulant based coherent signal subspace method for bearing and range estimation

In the previous section, a modified MUSIC algorithm has been proposed in order to estimate both the range and the bearing objects at a fixed frequency. In this section, the frequency diversity of wideband signals is considered. The received signals come from the reflections on the buried objects thus these signals are totally correlated and the MUSIC method looses its performances if any preprocessing is used before as the spatial smoothing (Pillai & Kwon, 1989) or the frequential smoothing (Frikel & Bourennane, 1996), (Valaee & Kabal, 1995). It appears clearly that it is necessary to apply any preprocessing to decorrelate the signals. According to the published results (Pillai & Kwon, 1989), the spatial smoothing needs a greater number of sensors than the frequential smoothing.

In this section, the employed signals are wideband. This choice is made in order to decorrelate the signals by means of an average of the focused spectral matrices. Therefore the objects can be localized even if the received signals are totally correlated. This would have not been possible with the narrowband signals without the spatial smoothing. Among the frequential smoothing based processing framework (Maheswara & Reddy, 1996), (Pillai & Kwon, 1989), we have chosen the optimal method which is the bilinear focusing operator (Frikel & Bourennane, 1996), (Valaee & Kabal, 1995), in order to obtain the coherent signal subspace. This technique divides the frequency band into L narrowbands (Frikel & Bourennane, 1996), (Valaee & Kabal, 1995), then, transforms the received signals in the L bands into the focusing frequency f_0. The average of the focused signals is then calculated and consequently decorrelates the signals (Hung & Kaveh, 1988), (Wang & Kaveh, 1985). Here, f_0 is the center frequency of the spectrum of the received signal and it is chosen as the focusing frequency. The following is the step-by-step description of the technique:

1. use an ordinary beamformer to find an initial estimate of P, θ_k and ρ_k for $k = 1, ..., P$,

2. fill the transfer matrix,

$$\hat{\mathbf{A}}(f_n) = [\mathbf{a}(f_n, \theta_1, \rho_1), \mathbf{a}(f_n, \theta_2, \rho_2), ..., \mathbf{a}(f_n, \theta_P, \rho_P)], \tag{24}$$

where each component of the directional vector $\mathbf{a}(f_n, \theta_k, \rho_k)$ for $k = 1, ..., P$, is filled using equation (19) or (20) considering the object shape,

3. estimate the cumulant slice matrix output sensors data $\mathbf{C}_1(f_n)$ at frequency f_n,
4. calculate object cumulant matrix at each frequency f_n using equation (10):

$$\mathbf{U}_s(f_n) = (\hat{\mathbf{A}}^+(f_n)\hat{\mathbf{A}}(f_n))^{-1}\hat{\mathbf{A}}^+(f_n)[\mathbf{C}_1(f_n)]\hat{\mathbf{A}}(f_n)(\hat{\mathbf{A}}^+(f_n)\hat{\mathbf{A}}(f_n))^{-1}, \tag{25}$$

5. calculate the average of the cumulant matrices associated to the objects,

$$\bar{\mathbf{U}}_s(f_0) = \frac{1}{L}\sum_{n=1}^{L}\mathbf{U}_s(f_n), \tag{26}$$

6. calculate $\hat{\mathbf{C}}_1(f_0) = \hat{\mathbf{A}}(f_0)\bar{\mathbf{U}}_s(f_0)\hat{\mathbf{A}}^+(f_0)$
7. form the focusing operator using the eigenvectors,

$$\mathbf{T}(f_0, f_n) = \hat{\mathbf{V}}_s(f_0)\mathbf{V}_s^+(f_n) \tag{27}$$

where $\mathbf{V}_s(f_n)$ and $\hat{\mathbf{V}}_s(f_0)$ are the eigenvectors associated with the largest eigenvalues of the cumulant matrix $\mathbf{C}_1(f_n)$ and $\hat{\mathbf{C}}_1(f_0)$, respectively.

8. form the average slice cumulant matrix $\bar{\mathbf{C}}_1(f_0)$ and perform its eigendecomposition,

$$\bar{\mathbf{C}}_1(f_0) = \frac{1}{L}\sum_{n=1}^{L}\mathbf{T}(f_0, f_n)\mathbf{C}_1(f_n)\mathbf{T}^+(f_0, f_n) \tag{28}$$

The modified spatial spectrum of MUSIC method for wideband correlated signals is given by

$$Z_{wb}(f_0, \theta_k, \rho_k) = \frac{1}{\mathbf{a}^+(f_0, \theta_k, \rho_k)\bar{\mathbf{V}}_b(f_0)\bar{\mathbf{V}}_b^+(f_0)\mathbf{a}(f_0, \theta_k, \rho_k)}, \tag{29}$$

where $\bar{\mathbf{V}}_b(f_0)$ is the eigenvector matrix of $\bar{\mathbf{C}}_1(f_0)$ associated to the smallest eigenvalues.

5. Experimental setup

An underwater acoustic data have been recorded in an experimental water tank (figure 2) in order to evaluate the performances of the developed method. This tank is fill of water and homogeneous fine sand, where are buried three cylinder couples and one sphere couple, full of water or air, between 0 and 0.005 m, of different dimensions (see table 1). The considered sand has geoacoustic characteristics near to those of water. Consequently, we can make the assumption that the objects are in a free space. The experimental setup is shown in figure 3 where all the dimensions are given in meter. The considered cylindrical and spherical shells are made of dural aluminum with density $D_2 = 1800\ \mathrm{kg/m^3}$, the longitudinal and transverse-elastic wave velocities inside the shell medium are $c_l = 6300\ \mathrm{m/s}$ and $c_t = 3200$

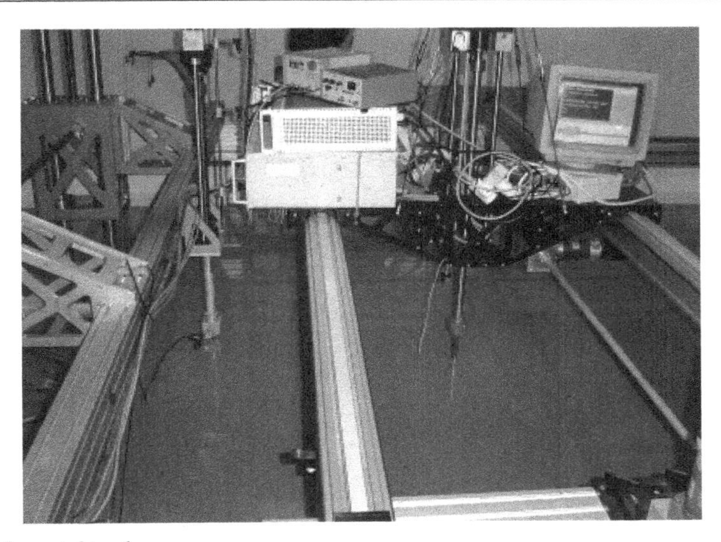

Fig. 2. Experimental tank

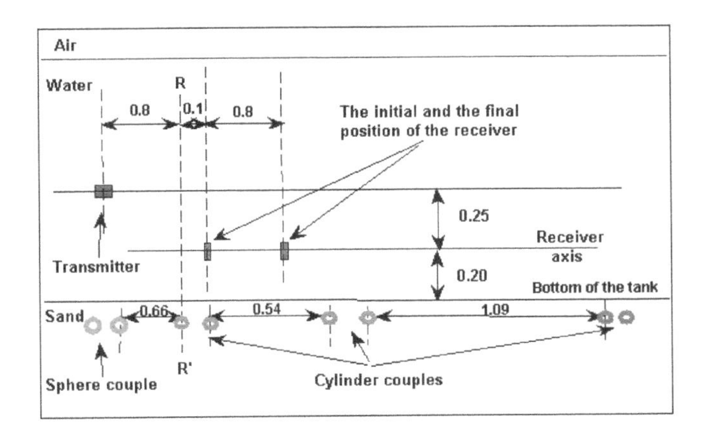

Fig. 3. Experimental setup

m/s, respectively. The external fluid is water with density $D_1 = 1000$ kg/m^3 and the the internal fluid is water or air with density $D_{3air} = 1.2\ 10^{-6}$ kg/m^3 or $D_{3water} = 1000$ kg/m^3. We have done eight experiments where the transmitter is fixed at an incident angle $\theta_{inc} = 60°$ and the receiver moves horizontally, step by step, from the initial to the final position with a step size $d = 0.002$ m and takes 10 positions in order to form an array of sensors with $N = 10$. The distance, between the transmitter, the RR' axis and the receiver, remains the same. First time, we have fixed the receiver horizontal axis at 0.2 m from the bottom of the tank, then, we have done four experiments, Exp. 1, Exp. 2, Exp. 3 and Exp. 4, associated to the following

	1st couple	2nd couple
Outer radius α_k (m)	$\alpha_{1,2} = 0.30$	$\alpha_{3,4} = 0.01$
Filled of	air	air
Separated by (m)	0.33	0.13

	3rd couple	4th couple
Outer radius α_k (m)	$\alpha_{5,6} = 0.018$	$\alpha_{7,8} = 0.02$
Filled of	water	air
Separated by (m)	0.16	0.06

Table 1. characteristics of the various objects (the inner radius $\beta_k = \alpha_k - 0.001$ m, for $k = 1, ..., P$)

configuration; the RR' axis is positioned on the sphere couple, the 1st, the 2nd and the 3rd cylinder couple, respectively. Second time, the receiver horizontal axis is fixed at 0.4 m and in the same manner we have done four other experiments Exp. 5, Exp. 6, Exp. 7 and Exp. 8 associated to each position of the RR'. Thus, for each experiment, only one object couple is radiated by the transmitter, where the transmitted signal has the following properties; impulse duration is 15 μs, the frequency band is $[f_L = 150, f_U = 250]$ kHz, the mid-band frequency is $f_0 = 200$ kHz and the sampling rate is 2 MHz. The duration of the received signal is 700 μs.

5.1 Experimental data

At each sensor, time-domain data is collected and the typical sensor output signals corresponding to one experiment are shown in figure 4. The power spectral density of the sensor output signal is presented in figure 5.

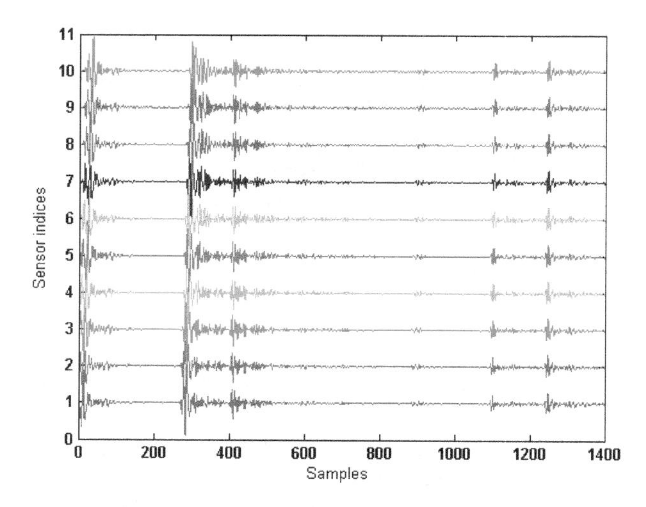

Fig. 4. Observed sensor output signals

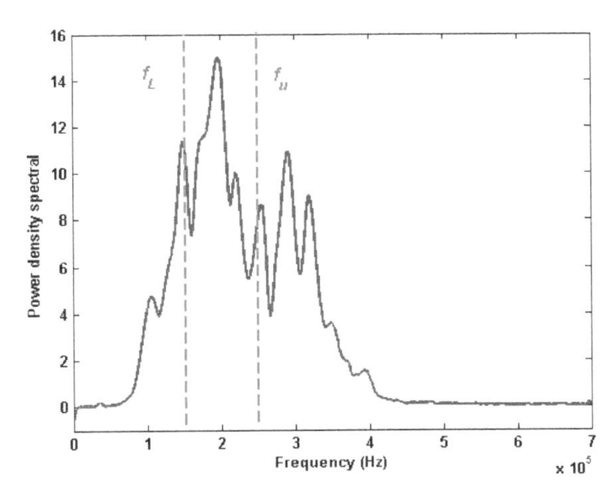

Fig. 5. Power spectral density of sensor output signal.

6. Results and discussion

The steps listed above in section 4, are applied on each experimental data, thus, an initialization of θ, ρ and P has been done using the conventional beamformer and for example for Exp. 1, those three parameters have been initialized by $P = 1$, $\theta_1 = 15^{\circ}$ and $\rho_1 = 0.28$ m. Moreover, the average of the focused slice cumulant matrices is calculated using $L = 50$ frequencies chosen in the frequency band of interest $[f_L, f_U]$. Moreover, a sweeping on the bearing and the range have been applied ($[-90^{\circ}, 90^{\circ}]$ for the bearing with a step 0.1° and $[0.2, 1.5]$ m for the range with a step 0.002 m). The obtained spatial spectrum of the modified MUSIC method are shown in figures 6 to 13. On each figure, we should have two peaks associated to one object couple and that is what appears on the majority of figures.

Table 2 summarizes the expected and the estimated range and bearing objects obtained using the MUSIC algorithm alone, then using the modified MUSIC algorithm with and without a frequential smoothing. The indexes 1 and 2 are the 1st and the 2nd object of each couple of cylinders or spheres. Note that the obtained bearing objects after applying the conventional MUSIC algorithm are not exploitable. Similar results are obtained when we apply the modified MUSIC algorithm without a frequential smoothing, because the received signals are correlated due to the small distance that separate the objects each other and the use of a single transmitter sensor. However, satisfying results are obtained when we apply the modified MUSIC algorithm with a frequential smoothing, thus the majority of bearing and range objects are successfully estimated. Furthermore, the difference between the estimated value $(\theta_{(1,2)est}, \rho_{(1,2)est})$ and the expected value $(\theta_{(1,2)exp}, \rho_{(1,2)exp})$ is very small and only two cylinders were not detected in Exp. 6 and Exp. 8, because, the received echoes, associated to these cylinders, are rather weak. Thus, it is important to realize that there is some phenomenons which complicate the object detection in experimental tank, for example, the attenuation of high frequencies in sediment is much higher than the low frequencies and due to the small dimensions of the experimental tank, the frequencies used here are $[150, 250]$ kHz which represent high frequencies.

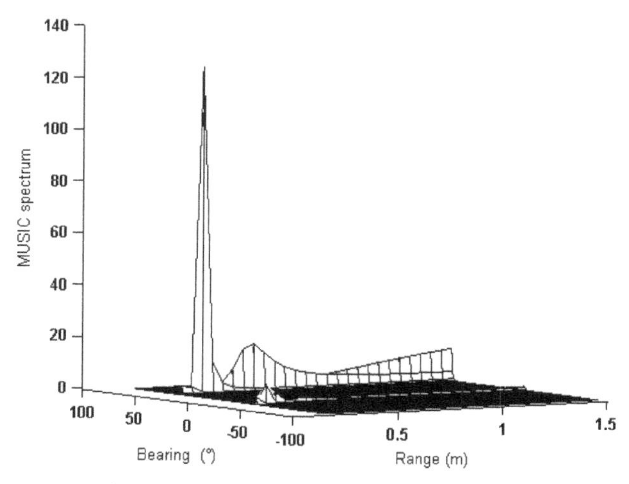

Fig. 6. Spatial spectrum of the developed method for air sphere couple (Exp.1.).

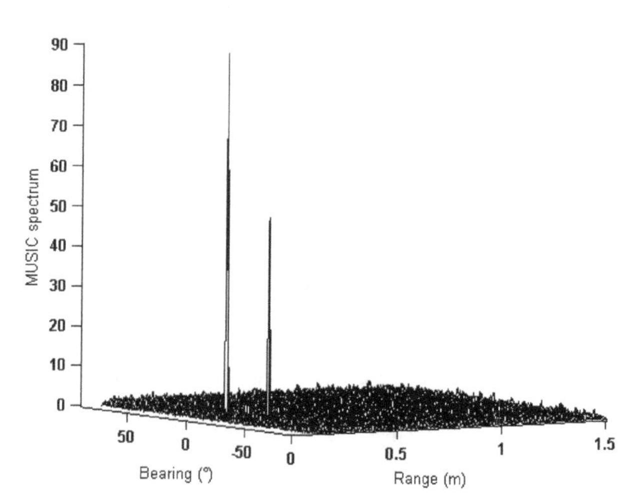

Fig. 7. Spatial spectrum of the developed method for small air cylinders couple (Exp.2).

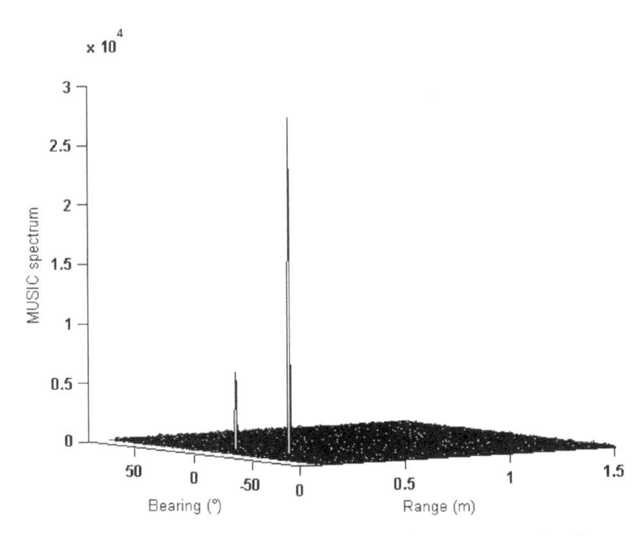

Fig. 8. Spatial spectrum of the developed method for big water cylinders couple (Exp.3).

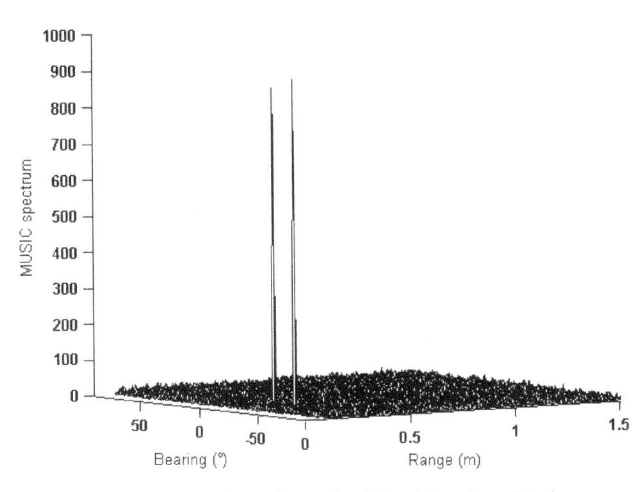

Fig. 9. Spatial spectrum of the developed method for big air cylinders couple (Exp.4.).

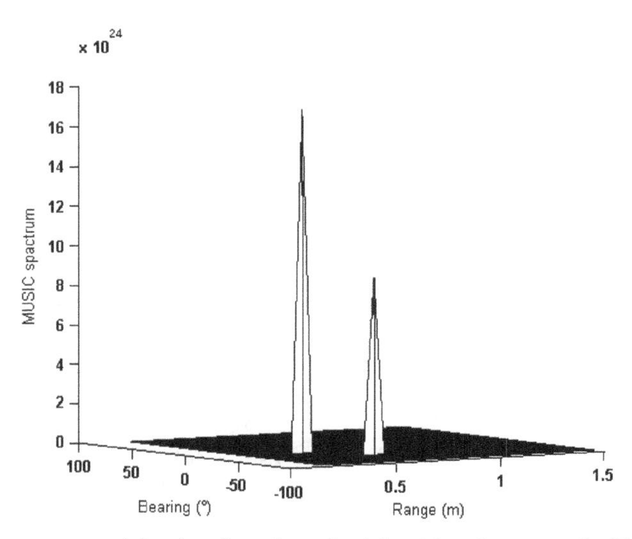

Fig. 10. Spatial spectrum of the developed method for Air sphere couple (Exp.5.).

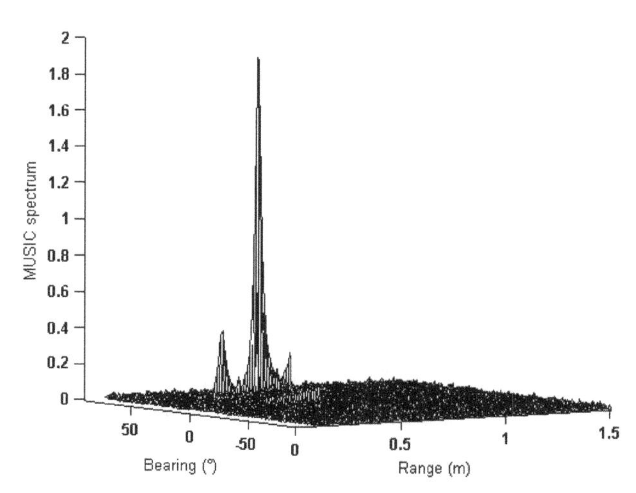

Fig. 11. Spatial spectrum of the developed method for small air cylinders couple (Exp.6).

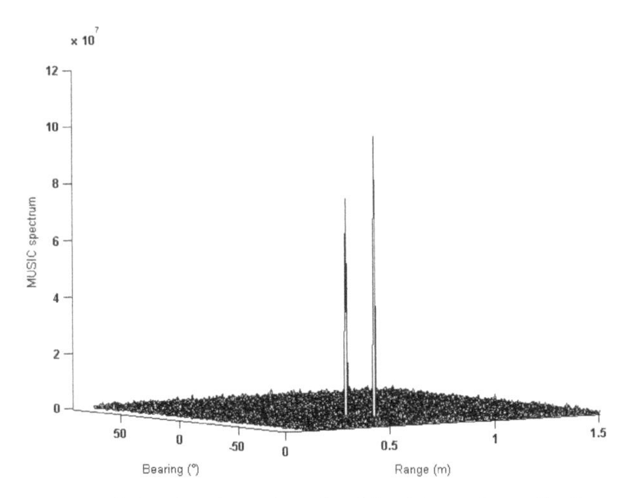

Fig. 12. Spatial spectrum of the developed method for big water cylinders couple (Exp.7).

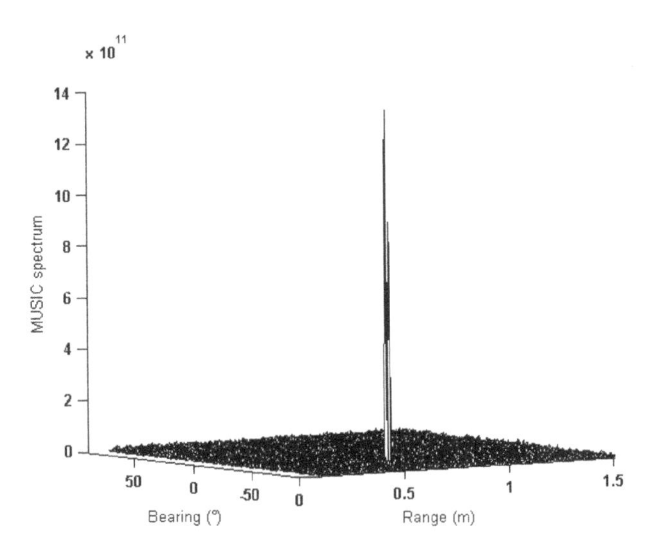

Fig. 13. Spatial spectrum of the developed method for big air cylinders couple (Exp.8.).

	Exp.1	Exp.2	Exp.3	Exp.4
$\theta_{1exp}(^\circ)$	−26.5	−23	−33.2	−32.4
$\rho_{1exp}(\mathrm{m})$	0.24	0.24	0.26	0.26
$\theta_{2exp}(^\circ)$	44	9.2	−20	5.8
$\rho_{2exp}(\mathrm{m})$	0.31	0.22	0.24	0.22
MUSIC				
$\theta_{1est}(^\circ)$	−18	−30	−40	−22
$\theta_{2est}(^\circ)$	30	−38	−48	−32
MUSIC NB				
$\theta_{1,2est}(^\circ)$	15	−12	−28	−10
$\rho_{1,2est}(\mathrm{m})$	0.28	0.23	0.25	0.24
MUSIC WB				
$\theta_{1est}(^\circ)$	−26	−23	−33	−32
$\rho_{1est}(\mathrm{m})$	0.22	0.25	0.29	0.28
$\theta_{2est}(^\circ)$	43	9	−20	6
$\rho_{2est}(\mathrm{m})$	0.34	0.25	0.25	0.23

	Exp.5	Exp.6	Exp.7	Exp.8
$\theta_{1exp}(^\circ)$	−50	−52.1	−70	−51.6
$\rho_{1exp}(\mathrm{m})$	0.65	0.65	1.24	0.65
$\theta_{2exp}(^\circ)$	−22	−41	−65.3	−49
$\rho_{2exp}(\mathrm{m})$	0.45	0.56	1.17	0.64
MUSIC				
$\theta_{1est}(^\circ)$	−58	25	−40	–
$\theta_{2est}(^\circ)$	−12	−40	−45	–
MUSIC NB				
$\theta_{1,2est}(^\circ)$	−35	−45	−70	−50
$\rho_{1,2est}(\mathrm{m})$	0.52	0.63	1.2	0.65
MUSIC WB				
$\theta_{1est}(^\circ)$	−49	−52	−70	−52
$\rho_{1est}(\mathrm{m})$	0.65	0.63	1.21	0.63
$\theta_{2est}(^\circ)$	−22	–	−65	–
$\rho_{2est}(\mathrm{m})$	0.44	–	1.2	–

Table 2. The expected (exp) and estimated (est) values of range and bearing objects. (negative bearing is clockwise from the vertical), NB: Narrowband, WB: Wideband)

7. Conclusion

In this chapter we have proposed a new method to estimate both the bearing and the range of buried objects in a noisy environment and in presence of correlated signals. To cope with the noise problem we have used high order statistics, thus we have formed the slice cumulant matrices at each frequency bin composed of clean data. Then, we have applied the coherent subspace method which consists in a frequential smoothing in order to cope with the signal correlation problem and to form the focusing slice cumulant matrix. To estimate the range and the bearing objects, the focusing slice cumulant matrix is used instead of using the spectral matrix and the exact solution of the acoustic scattered field is used instead of the plane wave model, in the MUSIC method. We considered objects with known shapes as cylindrical or spherical shells, buried in an homogeneous sand. Our method can be applied when the objects are located in the nearfield and rather in the farfield region of the sensor array. The performances of this method are investigated through real data associated to many spherical and cylindrical shells buried under the sand. The proposed method is superior in terms of performance to the conventional method. The range and the bearing objects are estimated with a significantly good accuracy due to the free space assumption.

8. Acknowledgment

The authors would like to thank Dr Jean-Pierre SESSAREGO for providing real data and Dr Zineb MEHEL-SAIDI for her useful collaboration.

9. References

Granara, M.; Pescetto, A.; Repetto, F.; Tacconi, G. & Trucco, A. (1998). Statistical and neural techniques to buried object detection and classification. *Proceedings of OCEANS'98 Conference*, Vol. 3, pp. 1269-1273, Nice France, oct. 1998.

Nicq, G. & Brussieux, M. (1998). A time-frequency method for classifying objects at low frequencies. *Proc. OCEANS'98 conference*, Vol. 1, pp. 148-152, Nice France, oct. 1998.

Guillermin, R.; Lasaygues, P.; Sessarego, J.P. & Wirgin, A. (2000). Characterization of buried objects by a discritized domain integral equation inversion method using born approximation. *Proceedings of 5th Eur. Conf. Underwater Acoustics*, Vol. 2, pp. 863-868, Lyon France, july 2000.

Hetet, A.; Amate, M.; Zerr, B.; Legris, M.; Bellec, R.; Sabel, J.C. & Groen, J. (2004). SAS processing results for the detection of buried objects with a ship-mounted sonar. *Proceedings of the 7th Eur. Conf. Underwater Acoustics*, Delft Netherland, july 2004.

Roux, P. & Fink, M. (2000). Time-reversal in a waveguide: Study of the temporal and spatial focusing. *J. Acoust. Soc. Am.*, Vol. 107, No. 5, (2000) (2418-2429).

Gönen, E. & Mendel, J.M. (1997). Applications of cumulants to array processing - Part III : Blind beamforming for coherent signals. *IEEE Trans. on signal processing*,Vol. 45, No. 9, (1997) (2252-2264), ISSN 1053-587X.

Mendel, J.M. (1991). Tutorial on higher order statistics(spectra) in signal processing and system theory : Theoretical results and some applications. *Proceedings of the IEEE*, 1991, Vol. 79, No. 3, pp. 278-305, march 1991.

Valaee, S. & Kabal, P. (1995). Wideband array processing using a two-sided correlation transformation. *IEEE Trans. on signal processing*, Vol.43, No.1, (jan. 1995)(160-172), ISSN 1053-587X .

Doolittle, R. & Uberall, H. (1966). Sound scattering by elastic cylindrical shells. *J. Acoust. Soc. Am.*, Vol. 39, No. 2, (1966) (272-275), ISSN 0001-4966.

Goodman, R.& Stern, R. (1962). Reflection and transmission of sound by elastic spherical shells. *J. Acoust. Soc. Am.*, Vol. 34, No. 3, (march 1962) (338-344), ISSN 0001-4966 .

Prada, C. & Fink, M. (1998). Separation of interfering acoustic scattered signals using the invariants of the time reversal operator. Application to Lamb waves characterization. *J. Acoust. Soc. Am.*, Vol. 104, No. 2, (august 1998).

Zhen, Y. (2001). Recent developments in underwater acoustics: Acoustic scattering from single and multiple bodies. *Proc. Natl.Sci. Counc. ROC(A)*, Vol.25, No.3, (2001)(137-150).

Lim,R.; Lopes, J. L.; Hackman, R. H. & Todoroff, D. G.(1993). Scattering by objects buried in underwater sediments: Theory and experiment. *J. Acoust. Soc. Am.*, Vol. 93, No. 4, (april 1993).

Tesei, A.; Maguer, A. & Fox, W. L. J. (2002). Measurements and modeling of acoustic scattering from partially and completely buried spherical shells. *J. Acoust. Soc. Am.*, Vol. 112, No. 5, (november 2002).

Junger, M. C.(1952). Sound scattering by thin elastic shells. *J. Acoust. Soc. Am.*, Vol. 24, No. 4, (july 1952).

Fawcett,J.A.; Fox, W.L. & Maguer, A. (1998). Modeling by scattering by objects on the seabed. *J. Acoust. Soc. Am.*, Vol. 104, No. 6, (dec. 1998) (3296-3304), ISSN 0001-4966.

Wang, H. & Kaveh, M. (1985). Coherent signal-subspace processing for the detection and estimation of angles of arrival of multiple wide band sources. *IEEE Trans. on Acoustics, Speech and Signal Processing*, Vol. 33, No. 4, (aug. 1985) (823-831), ISSN 0096-3518.

Yuen, N. & Friedlander, B. (1997). DOA estimation in multipath : an approach using fourth order cumulants. *IEEE. Trans. on Signal Processing*, Vol. 45, No. 5, (may 1997) (1253-1263), ISSN 1053-587X .

Bourennane, S. & Bendjama, A. (2002). Locating wide band acoustic sources using higher-order statistics. *Applied Acoustics*, Vol. 63, No. 3, (march 2002) (235-251).

Pillai, S.& Kwon, B. (1989). Forward/Backward spatial smoothing techniques for coherent signal identification. *IEEE Trans. on Acoustics, Speech and Signal Processing*, Vol. 37, No. 1, (jan. 1989)(8-15), ISSN 0096-3518 .

Frikel, M. & Bourennane, S. (1996). Fast algorithm for the wideband array processing using two-sided correlation transformation. *Proceedings of EUSIPCO'96*, Vol. 2, pp. 959-962, Trieste Italy, Sept. 1996.

Maheswara Reddy, K.& Reddy, V.U. (1996). Further results in spatial smoothing. *Signal processing*, Vol. 48, (1996) (217-224).

Hung, H. & Kaveh, M. (1988). Focusing matrices for coherent signal-subspace processing. *IEEE Trans. on Acoustics, Speech and Signal Processing*, Vol. 36, No. 8, (Aug. 1988) (1272-1281), ISSN 0096-3518.

3

Adaptive Technique for Underwater Acoustic Communication

Shen Xiaohong, Wang Haiyan, Zhang Yuzhi and Zhao Ruiqin
College of Marine Engineering, Northwestern Polytechnical University, Xi'an, China

1. Introduction

Compared with the electromagnetic wave channel, the UWA (Underwater Acoustic) channel is characterized by large transmission delay, transmission loss increased with distance and frequency, serious multi-path effect, and remarkable Doppler Effect. These characteristics greatly influence the performance of UWA communication, and restrict channel capacity. Performance of UWA communication is far away from telecommunication, even if employ the similar techniques. Especially, available bandwidth is limited by spread distance because transmission loss increases with longer range distance. For example, the bandwidth for 5 km is 10 kHz whereas the bandwidth for 80 km is only 500Hz. For constant data rate, the bandwidth is confined by maximum working range and worst channel environment, so the utilization ratio of channel is very low. Besides the reasonable modulation model and advanced signal processing method which improves the detection performance of the receiver, the most fundamental way for reliable communication is adaptive modulation which greatly improves utilization ratio of bandwidth.

Modulation and detection techniques used for UWA communication include phase coherent (PSK and QAM) and noncoherent (FSK) techniques. The choice of modulation mode is based on the UWA channel parameters, such as multipath and the Doppler spread, as well as the SNR. The spread factor of the UWA channel determines whether phase coherent communications are possible. If so, an equalizer is employed to combat any intersymbol interference. If the channel varies too rapidly, noncoherent signaling is chosen. The choice of modulation is determined by the operator.

The objective of this chapter is based on the analysis of characteristics of UWA channel, developing algorithms that can self-adapting select the best technique for time varying UWA channel. Then the dynamic modulation and bandwidth optimization for high-rate UWA communication are deduced. At last the simulation results are shown.

2. Fading characteristics of UWA channel

2.1 Characteristic of UWA channel

Due to the absorption of medium itself, the wavefront expansion in sound propagation, the bending of acoustic ray, the scattering caused by various kinds of nonuniformity in ocean and so on, the acoustic intensity will weaken in its propagation direction. The fading will change with distance, frequency and sensors location, which can be divided into large-scale fading and small-scale fading.

Large-scale fading: It is caused by sound propagation and sound ray convergence. The fading caused by sound propagation is composed by transmission loss and absorption loss, which is a function of range and frequency. It can be expressed as

$$TL = n \cdot 10 \log r + \alpha r \tag{1}$$

where n=1,2,or 1.5, r is the range of communication, α is absorption coefficeent, which is a function of signal frequency f, and its value can be given by Thorp experience formula:

$$\alpha(f) = \frac{0.11 f^2}{1+f^2} + \frac{44 f^2}{4100 + f^2} + 3.0 \times 10^{-4} f^2 + 3.3 \times 10^{-3} \text{ dB/km} \tag{2}$$

In the process of signal propagation, sound energy is strong in one area while weak in another area which is known as convergence fading.

Small-scale fading: small scale fading is composed by multipath and Doppler spread. The impulse response function is

$$h(\tau, t) = \sum_{i=1}^{p} a_i(t) \exp[j\phi_i(t)] \delta[\tau - \tau_i(t)] \tag{3}$$

where P is the number of multiple paths, α_i and ϕ_i are the amplitude and phase of i-th path respectively. τ_i is the time delay of the i-th path, and is uniformly distributed between $\tau_{\min}$ and $\tau_{\max}$, where $\tau_{\max}$ is maximum delay spread . This kind of spread in time domain is corresponding to frequency selective decline in frequency domain, and the relation between coherence bandwidth B_{coh} and maximum relative delay is $B_{cok} = 1/\tau_{\max}$. Multipath fading which caused by interference of multipath signals is related to the frequency of the signal, physical characteristics of the ocean, spatial location of the transmitter and the receiver.

Doppler Effect is caused by the relative motion between the transmitter and the receiver or the medium flow in UWA channel. Because multiple paths spread through different tracks, the received Doppler frequencies are also different from each other. Therefore the received signal is frequency spreading and the spread is measured by parameter $f_d = f\upsilon_{\max}/c$, where $\upsilon_{\max}$ is the maximum relative radial velocity of multiple paths, c is the propagation velocity of sound waves, and f is the frequency of the signal transmission. Doppler spread in frequency domain is corresponding to time selective decline in time domain. Time selective decline is measured by relative time $T_{cok} = 1/f_d$ .The larger the relative time is, the slower channel varies. Conversely, the smaller the relative time is, the quicker channel varies.

By the above analysis, when the sea area, positions of the transmitter and of the receiver are established, large-scale fading is only the function of distance and frequency, and small scale fading is a random function of distance and time. In the adaptive UWA communication, big scale fading is slow fading, and small scale fading is fast fading. The relations between range and fading are shown in Fig.1. With defined distance, the relations between time and fading are shown in Fig.2.

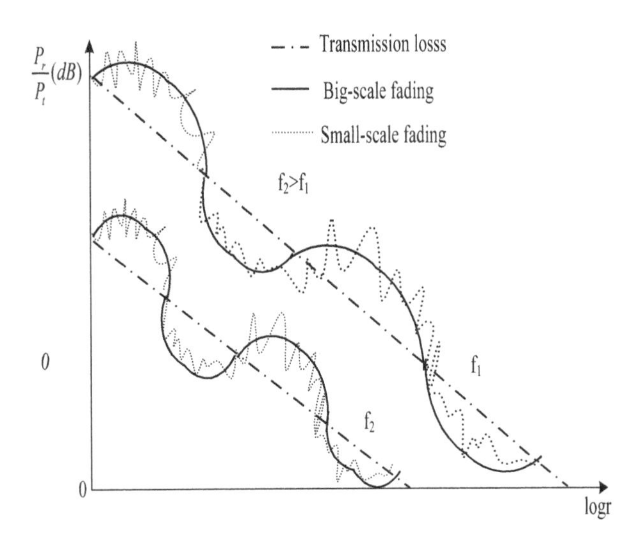

Fig. 1. The relations between range and fading

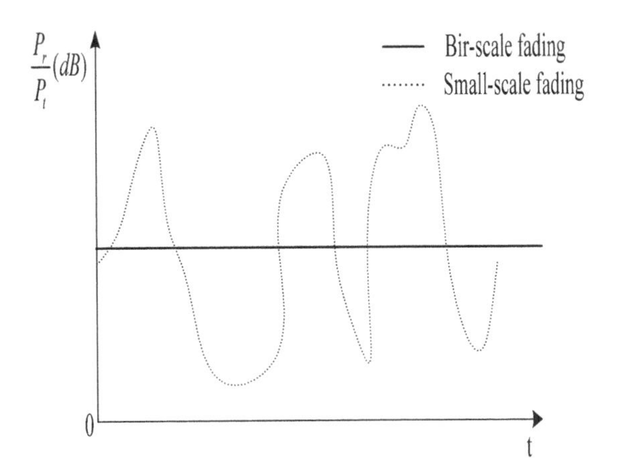

Fig. 2. The relations between time and fading

2.2 The relationship between bandwidth and frequency/ range in UWA channel

If only considering the large-scale fading, according to sonar function, the SNR(signal-to-noise ratio) of receiver will be

$$\mathrm{SNR} = SL - TL - NL - 10\log \mathrm{B} \tag{4}$$

where SL is sound source level (dB), NL is noise spectrum level, B is the bandwidth. Assuming NL=45dB, the relationship between SNR of receiver and frequency (1~30kHz),

SNR of receiver and range(1~100km) are shown in Fig.3. From the figure we can see, if the transmitting power and the SNR of receiver are defined, the system bandwidth is the function of distance and frequency.

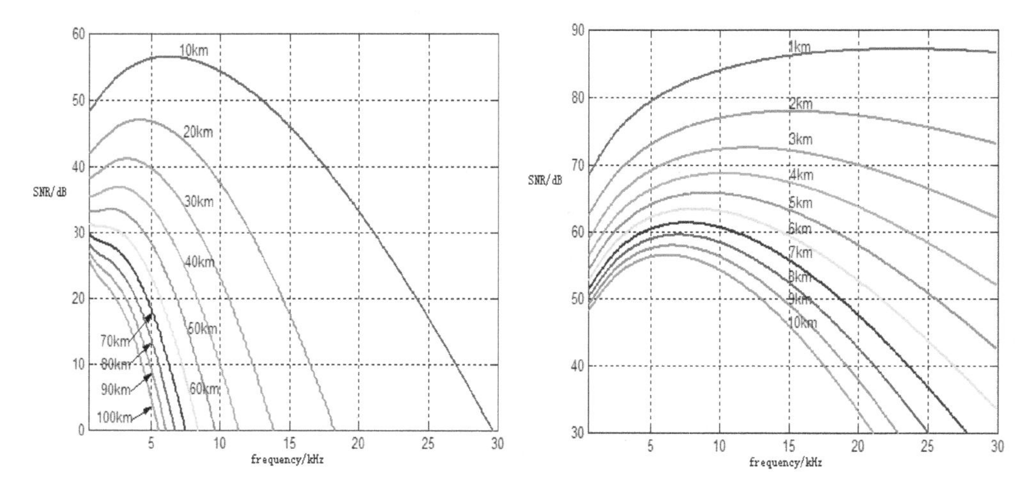

Fig. 3. The relationship between SNR and frequency, SNR and range

3. Channel capacity of UWA channel

3.1 Channel capacity of AWGN (Additive White Gaussian Noise) time invariant channel

Assuming $n(t)$ is AWGN, $N_0/2$ is PSD (Power Spectral Density) of AWGN, $\overline{P}$ is the average transmitted power, B is the receiving bandwidth, C is the channel capacity per second(bit/s). When the channel gain is constant 1, the SNR of the receiver is constant $\gamma(t) = \overline{P}/(N_0 B)$, and then the channel capacity limited by average power can be expressed as

$$C = B\log_2(1+\gamma) \tag{5}$$

This formula shows the channel capacity is proportional to the bandwidth, and increases with the improvement of received SNR.

3.2 Channel capacity of time variant flat fading channel

Supposing the gain of stationary and ergodic channel is $\sqrt{g(t)}$, $g(t)$ obeys distribution p(g), and is unrelated with channel input. So the instantaneous received SNR is $\gamma(t) = g(t)\overline{P}/(N_0 B)$, the distribution of $\gamma(t)$ is determined by $g(t)$, and the channel capacity is

$$C = \int_0^{\infty} B\log_2(1+\gamma)p(\gamma)d\gamma \tag{6}$$

If the sending power varies with γ, the interrupt threshold γ_0 can be calculated by the formula $\int_{\gamma_0}^{\infty}(1/\gamma_0 - 1/\gamma)p(\gamma)d\gamma = 1$, it can be proved when the optimal power allocation is $\frac{P(\gamma)}{\overline{P}} = \begin{cases} 1/\gamma_0 - 1/\gamma & \gamma \geq \gamma_0 \\ 0 & \gamma < \gamma_0 \end{cases}$, channel capacity achieves maximum as

$$C = \int_{\gamma_0}^{\infty} B\log_2(\gamma/\gamma_0)p(\gamma)d\gamma \tag{7}$$

This capacity can be obtained by the time-varying transmission rates. Instantaneous SNR γ corresponds to the rate $B\log_2(\gamma/\gamma_0)$. This formula shows: If channel condition gets deteriorated, then the transmitter will reduce sending power and transmission rate. If the SNR falls below the interrupt threshold γ_0, the transmitter won't transmit signal. This method is called time domain Power Water-filling Allocation method.

3.3 Channel capacity of time invariant frequency selective fading channel

Supposing the channel gain *H(f)* is block fading, the whole bandwidth can be divided into many sub-channels whose bandwidth are B, and in each sub-channel $H(f) = H_i$ is constant. The SNR of i-th channel is $|H_i|^2 P_j/(N_0 B)$, where P_j is the allocated power when $\sum_j P_j < P$. *P* is the upper limit of the total power. The channel capacity is the sum of the whole sub-channels

$$C = \sum_{P_j:\sum_j P_j \leq P} B\log_2\left(1 + \frac{|H_j|^2 P_j}{N_0 B}\gamma\right) \tag{8}$$

It can be proved when the optimal power allocation is $\frac{P(\gamma)}{\overline{P}} = \begin{cases} 1/\gamma_0 - 1/\gamma & \gamma \geq \gamma_0 \\ 0 & \gamma < \gamma_0 \end{cases}$, channel capacity achieves maximum as

$$C = \int_{\gamma_0}^{\infty} B\log_2(\gamma/\gamma_0)p(\gamma)d\gamma \tag{9}$$

Interrupt threshold can be calculated by $\sum(1/\gamma_0 - 1/\gamma_j) = 1$. The capacity *C* is achieved by allocating different power and data rate for multiple sub-channels. When SNR of channel is admirable, distribute more power and use high transmission rate. If channel gets deteriorated, the transmitter will reduce sending power and transmission rate. If the SNR falls below the interrupt threshold γ_0, the transmitter won't transmit signal. This method is called frequency domain Power Water-filling Allocation method.

3.4 Capacity of UWA channel

In order to simplify the analysis, this chapter assumes that the response function of the UWA channel is *H (f, t)*, and the bandwidth *B* can be divided into several sub-bandwidths by coherent bandwidth B_{coh}. Each sub-channel is independent time variant flat fading channel, then for the j-th sub-channel, the response function is $H(f,t)=H_j(t)$. According to the allocated average power of each sub-channel, the capacity of every flat fading sub-channel can be deduced. For sub-channels are independent from each other, the total power in time and frequency domain is the sum of capacity of each narrowband flat fading sub-channel.

$$C=\max_{\{\overline{P}_j\}:\sum_j \overline{P}_j \le \overline{P}} \sum_j C_j\left(\overline{P}_j\right) \tag{10}$$

where $C_j\left(\overline{P}_j\right)$ is the capacity of the sub-channel whose mean power is P and bandwidth is B_{coh}. Bandwidth can be given by Eq. 6 From the two dimensions Water-filling Allocation method of time domain and frequency domain, when $\frac{P_j}{P}=\begin{cases}1/\gamma_0-1/\gamma_j & \gamma_j \ge \gamma_0 \\ 0 & \gamma_j < \gamma_0\end{cases}$, the channel capacity achieves maximum as

$$C=\sum_j \int_{\gamma_0}^{\infty} B_c \log_2\left(\gamma_j/\gamma_0\right) p\left(\gamma_j\right) d\gamma_j \tag{11}$$

Interrupt threshold γ_0 can be calculated by $\sum_j \int_{\gamma_0}^{\infty}\left(1/\gamma_0-1/\gamma_j\right)p\left(\gamma_j\right)d\gamma_j=1$, which is the same for each sub-channel.

To achieve the capacity in UWA communication, multi-carrier transmission should be used, and the power of each sub-channel is allocated by SNR. In one sub-channel, the transmission power and rate vary with the channel condition by Water-filling Allocation method.

4. Relationship between BER and SNR in digital communication

PSK(Phase Shift Keying), QAM(Quadrature Amplitude Modulation) and FSK (Frequency Shift Keying) are the common modulation models used in UWA communication. Supposing the transmission signal $s_T(t)$, then the received signal with AWGN is

$$s_r(t)=s_T(t)+n(t) \qquad (0 \le t \le T) \tag{12}$$

where *n(t)* is AWGN sample with power spectrum density $\Phi_{nn}(f)=\mathrm{N}_0/2$. *N0* is the average power spectrum density of AWGN. With the optimized receiver, symbol error rate, bit error rate and bandwidth ratio of several modulation models are shown in Table1. ε_b is signal power per bit.

According to Table1, when the BER is $P_b=10^{-6}$ or $P_b=10^{-4}$, the SNR that M-ary modulation needs to transmit 1bit information is shown in Table2. The interrupt threshold in UWA communication is defined by the SNR.

	SER	BER	bandwidth ratio
PSK	$P_B \approx 2Q\left(\sqrt{\frac{2k\varepsilon_b}{N_0}}\sin\frac{\pi}{M}\right)$	$P_b \approx \frac{1}{k}P_B$	$\frac{R_b}{W} = \frac{1}{2}\log_2 M$
QAM	$P_B \approx 4Q\left[\sqrt{\frac{3k\varepsilon_b}{(M-1)N_0}}\right]$	$P_b \approx \frac{1}{k}P_B$	$\frac{R_b}{W} = \begin{cases}\frac{1}{2}\log_2 M & M \le 4 \\ \log_2 M & M > 4\end{cases}$
FSK (non-coherent detection)	$P_B = \sum_{n=1}^{M-1}(-1)^{n+1}\binom{M-1}{n}\frac{1}{n+1}e^{\frac{-nk\varepsilon_b}{N_0(n+1)}}$	$P_b = \frac{2^{k-1}}{2^k - 1}P_B$	$\frac{R_b}{W} = \frac{\log_2 M}{M}$

Table 1. SER, BER and bandwidth ratio of M-ary modulation models

	Modulation mode	M=2	M=4	M=8	M=16	M=32
$P_b = 10^{-6}$	PSK	10.51	10.51	13.95	18.42	23.34
	QAM		10.51	13.25	14.39	16.59
	FSK	13.51	10.75	9.23	8.21	7.44
$P_b = 10^{-4}$	PSK	8.39	8.39	11.71	16.14	21.01
	QAM		8.39	11.28	12.19	14.41
	FSK	11.39	8.77	7.35	6.43	5.76

Table 2. SNR to transmit 1bit information of M-ary modulation (dB)

5. Adaptive UWA communication

The UWA communication is power limited communication because of cavitation phenomenon and field effect. The large-scale fading is slow fading, and the small-scale fading is fast fading.

From analysis of section 3.4, adaptive modulation should be used to optimize the bandwidth ratio. By the use of multi-carrier transmission, the power of each sub-channel is allocated by SNR of receiver to achieve dynamic bandwidth. When SNR of channel is high, the transmitter distributes more power, or otherwise distributes less power. If the SNR falls below the interrupt threshold γ_0, the transmiter will not transmit signal. In one sub-channel the transmission power and rate vary with the channel condition by Water-filling Allocation method. When SNR of channel is admirable, the transmitter distributes more power and uses higher transmission rate, or otherwise reduces sending power and transmission rate. If the SNR falls below the interrupt threshold γ_0, do not transmit signal. It is dynamic modulation in one sub-carrier.

5.1 Adaptive transmission system

In adaptive modulation system, at first, the transmitter sends a test signal to link the receiver, and then the receiver estimates the instantaneous channel characteristics and feeds back the updates to the transmitter, at last the transmitter selects a suitable modulation scheme and bandwidth to transmit information. The receiver estimates channel and feeds back the updates to transmitter at the end of a frame. The adaptive system process block diagram is shown in Fig.4. The estimated received SNR of j-th channel at i-th time is $\hat{\gamma}_j[i] = \overline{P}_j \hat{g}_j[i] / (N_0 B_j)$. $\hat{g}_j[i]$ is the estimation of power gain of j-th channel. The relevant parameters are regulated based on estimated value at the time of integral times of code period *Ts*.

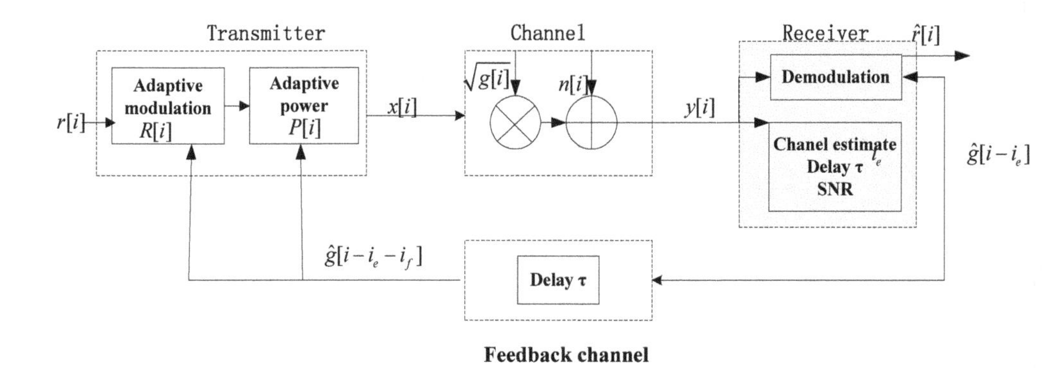

Fig. 4. Block diagram of adaptive system

5.2 The adaptability criterion of large-scale fading and small-scale fading

Supposing the transmitter sends information to the receiver at time i, after time delay i_e, the receiver receivs the estimated SNR $\hat{\gamma}_j[i]$ of j-th sub-channel at time i. Finally, with the additional feed back time delay i_f, the receiver gets the information after time delay $i_d = i_e + i_f$. Assuming the channel estimation and feed back information are both errorless, then the instantaneous BER is $P_b(\gamma) = P_b$, and supposing $\hat{g}[i - i_d]$ is the actual value, the total BER of QAM is

$$P_b(\gamma[i], \gamma[i - i_d]) \leq 0.2[5P_b]^{\varepsilon[i, i_d]} \tag{13}$$

It can be improved when Doppler frequency shift *fd* meets $i_d = i_e + i_f < 0.001 f_d$, the average BER approaches desired value P_b. That is to say, the rate of channel variation effects the updates rate and estimated error of $\gamma[i]$.

As mentioned above, there are two aspects affecting channel gain, large-scale fading and small-scale fading. When Doppler frequency shift is small, the large-scale fading is constant and the small-scale fading varies slowly. In another word, in short range, estimated error and feed back delay can not weaken the system. When Doppler frequency shift is large, small-scale fading varies severely, that is system can not estimate channel and feed back information effectively in long range. Different adaptive control methods should be adopted according to different range, and there is a definite relation between control methods and communication range. Supposing the distance is r, UWA velocity is 1500m/s, time delay $i_d \geq 2r/c$, when $d < 0.001cf_d/2$, system chooses the model to adapt small-scale fading, otherwise to adapt large-scale fading.

5.3 Dynamic modulation and bandwidth optimization

Adaptive dynamic modulation and bandwidth optimization are powerful techniques to improve the energy efficiency and bandwidth ratio over UWA fading channel. Adaptive frequency and bandwidth are natural choices over UWA fading channel. Adaptive modulation and power management over UWA fading channel is investigated in this section.

5.3.1 Define carrier frequency and bandwidth according to the range

For any UWA communication system, the most important parameters are carrier frequency and bandwidth, which are mainly dependent on range and ambient noise. Receiver expects a high SNR, so the principle to choose a proper carrier frequency and bandwidth is to maximize SNR. Based on sonar equation, the optimal frequency can be determined by

$$f_0 = \left(\frac{70.7}{r} \cdot \frac{d(FM)}{df} \right)^{1/2} \tag{14}$$

in which, *FM* is defined as *FM=SL-(NL-DI+DT)*, *SL* is the Source Level of transmitter， *NL* is water ambient Noise Level. The 3dB bandwidth of the optimistic frequency is the system bandwidth.

5.3.2 Define the modulation model and adaptive fading model

For double spread UWA channel, the spread factor is the product of time spread and frequency spread. If the spread factor accesses or overs 1, the channel will be overspread. Non-coherent modulation FSK has been considered as the only alternative for overspread channel.

According the estimated distance d and Doppler frequency fd, if $d < 0.001cf_d/2$, then the system will adapt to small-scale fading that is fast fading, otherwise will adapt to large-scale fading.

5.3.3 SNR estimation

SNR is an important parameter to optimize modulation and bandwidth in two dimensions channel. There are two estimation methods, one is estimating by sending the test training sequence, and the other is estimating by direct signals received. The former one is inefficient for lower bandwidth availability, and the later one is the main method of SNR estimation.

Supposing the transmission signal is $s_T(t)$, then the received signal through UWA channel with AWGN is

$$y(t)=\sum_{l=1}^{p}\alpha_l \exp[j(\phi_l(t)]s_T(t-\tau_l f_o)+n(t) \tag{15}$$

The self-correlation function of the received signal is

$$\begin{aligned}\phi_{yy}(\tau) &= E[y(t)y(t+\tau)] \\ &= E[y_I(t)y_I(t+\tau)]\cos 2\pi f_c\tau \\ &\quad -E[y_Q(t)y_I(t+\tau)]\sin 2\pi f_c\tau+\sigma_n^2\delta(\tau) \\ &= \phi_{ylyI}(\tau)\cos 2\pi f_c\tau-\phi_{ylyQ}(\tau)\sin 2\pi f_c\tau \ +\sigma_n^2\delta(\tau) \\ &= \sigma^2 J_0(2\pi f_m\tau)\cos(2\pi f_c\tau)+\sigma_n^2\delta(\tau)\end{aligned} \tag{16}$$

In the formula, σ^2 is signal power, σ_n^2 is noise power, $J_0(x)$ is the zero-order Bessel function of first kind, $f_m=\Delta f\ /\ f_0$ is normalization Doppler frequency. From the properties of channel impulse response, the above formula is discretized and normalized as

$$\varphi_{yy}(kT_s)=\frac{\sigma^2 J_0(2\pi f_m kT_s)\cos(2\pi f_c kT_s)}{\sigma^2+\sigma_n^2} \qquad (k=1,2,3...) \tag{17}$$

According to the definition of SNR

$$SNR=\frac{\varphi_{yy}(kT_s)}{J_o(2\pi f_m kT_s)\cos(2\pi f_c kT_s)-\varphi_{yy}(kT_s)} \qquad (k=1,2,3...) \tag{18}$$

If f_m is unknown, the equation set can be deduced from Eq.18

$$\left\{\begin{aligned}\gamma &= \frac{\varphi_{yy}(T_s)}{J_0(2\pi f_m T_s)\cos(2\pi f_c T_s)-\varphi_{yy}(T_s)} \\ \gamma &= \frac{\varphi_{yy}(2T_s)}{J_0(4\pi f_m T_s)\cos(4\pi f_c T_s)-\varphi_{yy}(2T_s)}\end{aligned}\right\} \tag{19}$$

γ can be calculated by the equation set. If the system adapts to slow fading channel, the estimated γ can be averaged.

5.3.4 Allocate the instantaneous power of each sub-channel based on the estimated SNR

First, the power of each sub-channel is allocated by $\frac{P_j}{P}=\begin{cases}1/\gamma_0-1/\gamma_j & \gamma_j\ge\gamma_0 \\ 0 & \gamma_j<\gamma_0\end{cases}$. Then, instantaneous power of each sub-channel is allocated by $\frac{P(\gamma)}{\overline{P}}=\begin{cases}1/\gamma_0-1/\gamma & \gamma\ge\gamma_0 \\ 0 & \gamma<\gamma_0\end{cases}$

5.4 The adaptive multi-carrier modulation in UWA communication

OFDM transports a signal-input data stream on several carriers within the usable frequency band of the channel. This is accomplished by partitioning the entire channel into N parallels, ideally orthogonal, and spectrally flat subchannels, each of equal bandwidth, and with center frequency f_n, $n=1...N$. Thus an OFDM symbol is consisted of several subcarriers which are modulated as PSK or QAM. Each subcarrier can be independently modulated in adaptive modulation schemes or all subcarriers may be modulated in same manner. The bandpass OFDM symbol can be expressed as follows:

$$x(t) = \sum_{n=0}^{N-1} d(n) e^{j2\pi f_0 t}, \quad t \in [t_0, t_0 + T_s] \tag{20}$$

where T_s is the symbol period, f_0 is the carrier frequency, N is the number of subcarriers, $d(n)$ is the PSK or QAM symbol, and t_0 is the symbol of starting time. The bandpass signal in Eq. 20 can also be expressed in the form of

$$x(t) = [\sum_{n=0}^{N-1} d(n) \exp(j\frac{2\pi}{T_s} nt)] \exp(j2\pi f_0 t) = X(t) \times \exp(j2\pi f_0 t) \tag{21}$$

The baseband signal $x(t)$ is sampled at a rate of f_s, then $t_k=k/f_s$. The baseband signal $X(t)$ can be expressed in the form of

$$X(k) = \sum_{n=0}^{N-1} d(n) \exp(j\frac{2\pi}{N} nk), \quad 0 \le k \le (N-1) \tag{22}$$

where $x(k)=x(t_k)$. Eq. 22 shows that $x(t)$ is the Inverse Fast Fourier Transform (IFFT) of $d(n)$.

At receiver the FFT is applied to the discrete time OFDM signal $x(t)$ to recovery the $d(n)$ symbols written in

$$d(n) = \sum_{k=0}^{N-1} X(k) \exp(-j\frac{2\pi}{N} nk), \quad 0 \le n \le (N-1) \tag{23}$$

Noncoherent detection MFSK transports a signal-input data stream on selected carriers among N parallel channels based on coding. A MFSK symbol can be considered as a special case of OFDM symbols. The condition of $d(n)$ to be satisfied is that

$$d(n) = \begin{cases} 1, & \text{if sending subcarrier} \\ 0, & \text{others} \end{cases} \tag{24}$$

It is clear from Eq. 22, 23 and 24 that dynamic adaptive modulation or demodulation combines MFSK and OFDM effectively and employs the FFT/IFFT algorithm to synthesize the modulation or demodulation without any other algorithms. It selects a suitable modulation or demodulation between OFDM and MFSK according to the estimated parameters of UWA communication channel.

6. Simulation and experiment

6.1 SNR estimation

In order to evaluate the performance of the dynamic adaptive system, an UWA communication experiment was conducted in a lake. The impulse response of the lake is shown in Fig.5.

According to the SNR estimated method, the 2FSK, 2PSK and 4QAM signals in different SNRs are respectively simulated in 100 times. Considering the SNR varying on the receiver and practical application, with the range of SNR [-10 dB, 10 dB], the estimated standard-deviations are shown in Fig.6

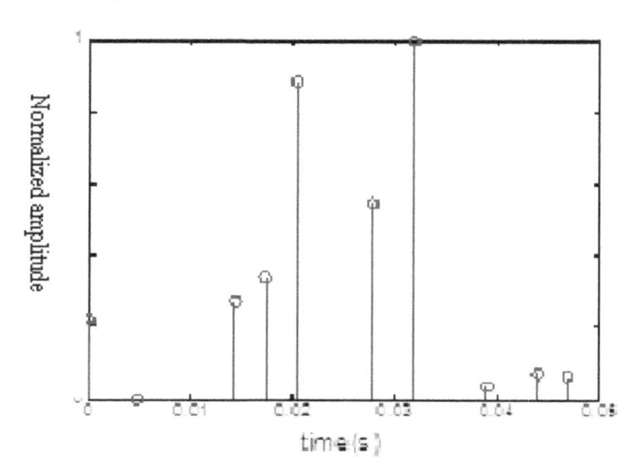

Fig. 5. Impulse response of the lake

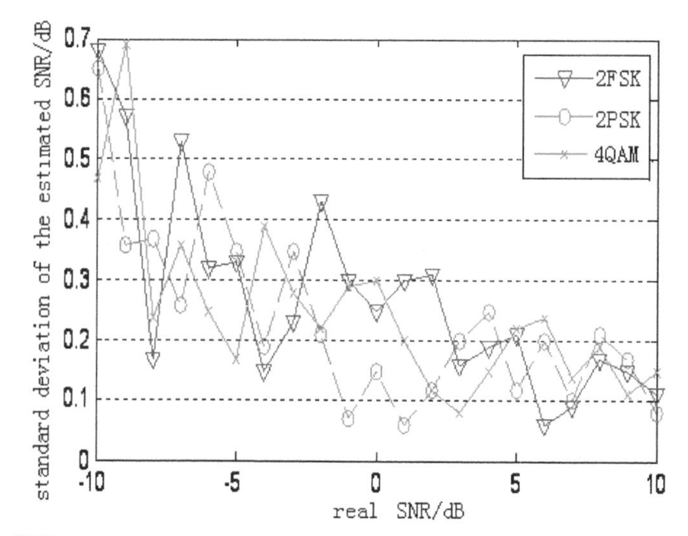

Fig. 6. Average SNR

From the figure, we can see the SNR estimated algorithm performance well. The standard deviation of estimated SNR is less than 0.7dB. So the estimation is available in adaptive UWA communication.

6.2 Simulations

The parameters of adaptive UWA communication system used for performance simulation are showed in Table 3. A linear frequency modulation signal is used for synchronization. The UWA channel is based on the Fig.5, and the relative velocity is 1.5m/s.

Modulation	Bandpass	Bandwidth	Guard interval	FFT /IFFT	Number of sub carrier
MFSK	2~2.2kHz	200Hz	0s	0.2s	1
	2~2.4kHz	400Hz			2
	2~2.8kHz	800Hz			4
	2~3.6kHz	1600Hz			8
OFDM	2~5 kHz	3 kHz	0.05s		300
	2~6 kHz	4 kHz			400
	2~7 kHz	5 kHz			500
	2~8 kHz	6kHz			600

Table 3. Adaptive UWA parameters

To overcome the ISI, system uses a grouped FSK modulation technique. The band is divided into 2 groups, only one of which is transmitted at one time. System uses multiple FSK modulation technique in a group. The band is divided into 2, 4 or 8 subbands based on bandwidth, in each of which an 8FSK signal is transmitted.

Fig. 7 and Fig.8 show that the bandwidth raises at the price of increasing SNR at the same error bit. Compared with Fig.3, 200Hz bandwidth 8FSK can be used for UWA communication at 100km range, and 1200Hz can be used for 40 km. OFDM with 3 kHz bandwidth can be used for UWA communication at 15 km range, and 6 kHz can be used for the short range.

6.3 Lake experiments

An UWA communication experiment which using adaptive system with 1/2 rate turbo error control coding was taken in lake. The results demonstrate the performance of adaptive communication system at the range from 5km to 25km. The modulation schemes, the number of subcarriers, data rate and the bit error rate are given in Table 4.

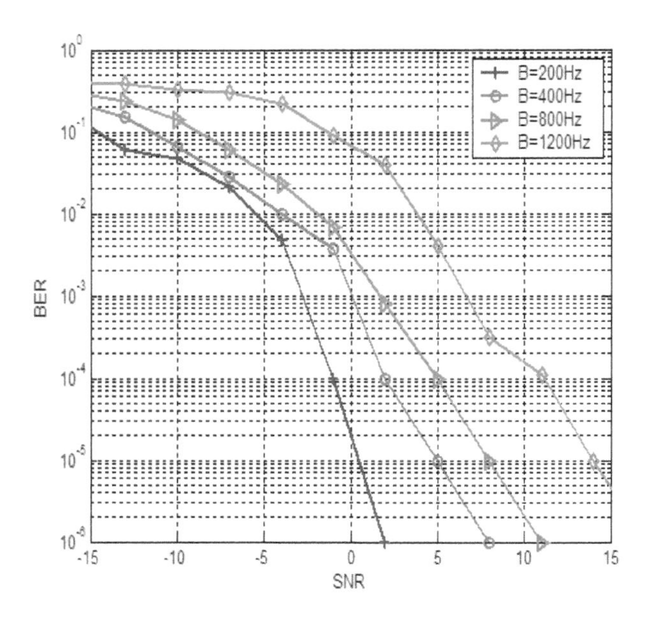

Fig. 7. BER Performance of four kinds of bandwidth of MFSK

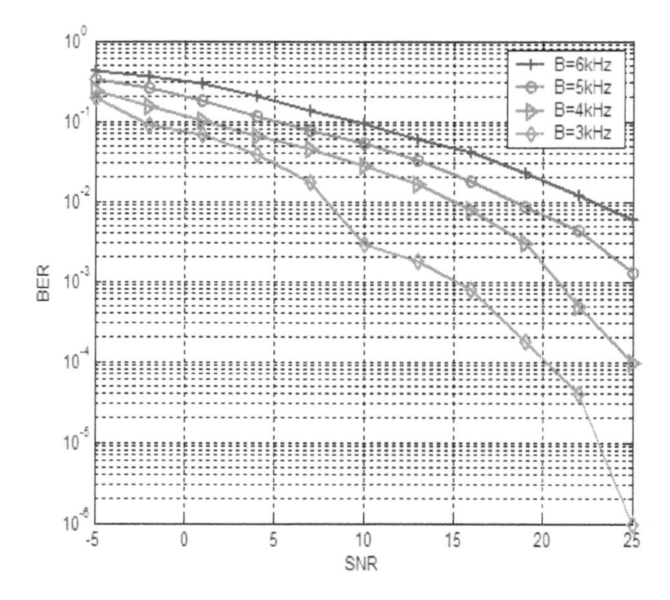

Fig. 8. BER Performance of four kinds of bandwidth of OFMD

Range /km	modulation	number of subcarriers	data rate bs-1	BER /%	BER after decoding
5	OFDM	1000	9090	5.33	$<10^{-4}$
10	OFDM	500	3400	8.40	$<10^{-4}$
15	OFDM	200	1360	8.32	$<10^{-4}$
25	8FSK	16	640	2.55	$<10^{-4}$

Table 4. Results of lake experiment

7. Conclusion

MFSK was seen as intrinsically robust for the time and frequency spreading of long range UWA channel. OFDM has been used in UWA communication at short or medium range. Adaptive UWA communication system combines MFSK and OFDM effectively, which dynamic selects modulation schema and optimizes bandwidth based on UWA communication estimation. This method has obvious advantages: being realized by DFT based filter banks as OFDM, good performance and the high frequency band efficiency in time varying fading UWA channel. Based on the results of simulation and experiments in a lake, it is shown that the adaptive UWA communication system is more efficient for high rate UWA communication not only at short range, but also at medium and long range.

8. References

Andrea Goldsmith. Wireless communications. Camgridge university press. 2005.

Benson A., Proakis J., Stojanovic M., Towards robust adaptive acoustic communications[C]. OCEANS 2000 MTS/IEEE Conference and Exhibition. 2:1243 ~ 1249 ,2000.

Bayan S. Sharif, Oliver R. Hinton, & Alan E. Adams. A Computationally Efficient Doppler Compensation System for Underwater Acoustic Communications. IEEE Journal of Oceanic Engineering, 2000; 25(1): 52-61.

G.Lapierre, N.Beuzelin.1995-2005:Ten years of active research on underwater acoustic communications in Brest. OCEANS 2005:425~430.

Kilfoyle Daniel B., Baggeroer Arthur B. The State of the Art in Underwater Acoustic Telemetry. IEEE Journal of Oceanic Engineering, 2000; 25(1): 4-27

K. F. Scussel, J. A. Rice, and S. Merriam. A new MFSK acoustic modem for operation in adverse underwater channels. Oceans'97, Halifax, NS, Canada, 1997.

Michele Zorzi. Energy-Efficient Routing Schemes for Underwater Acoustic Networks[J]. IEEE Trans. Commu.. 26(9):1754~1766,2008.

Stojanovic M. Recent Advances in High-speed Underwater Acoustic Communications. IEEE Journal of Oceanic Engineering, 1996; 21(2): 125-136.

Woodward B, Sari H. Digital UnderwaterAcoustic Voice Communications. IEEE Journal of Oceanic Engineering, 1996; 21(2):181-191.

Wang Haiyan, Jiang Zhe, Modifying SNR-Independent Velocity Estimation Method to Make it Suitable for SNR Estimation in Shallow Water Acoustic Communication. Journal of Northwestern Polytechnical University. 27(3):368~371,2009.

Yeung Lam F.,Robin S. et al Underwater acoustic modem using multi-carrier modulation OCEANS'2003, 2003;3:1368 - 137.

4

Array Processing: Underwater Acoustic Source Localization

Salah Bourennane, Caroline Fossati and Julien Marot
Institut Fresnel, Ecole Centrale Marseille
France

1. Introduction

Array processing is used in diverse areas such as radar, sonar, communications and seismic exploration. Usually the parameters of interest are the directions of arrival of the radiating sources. The High-Resolution subspace-based methods for direction-of-arrival (DOA) estimation have been a topic of great interest. The subspace-based methods well-developed so far require a fundamental assumption, which is that the background noise is uncorrelated from sensor to sensor, or known to within a multiplicative scalar. In practice this assumption is rarely fulfilled and the noise received by the array may be a combination of multiple noise sources such as flow noise, traffic noise, or ambient noise, which is often correlated along the array (Reilly & Wong, 1992; Wu & Wong, 1994). However, the spatial noise is estimated by measuring the spectrum of the received data when no signal is present. The data for parameter estimation is then pre-whitened using the measured noise. The problem with this method is that the actual noise covariance matrix varies as a function of time in many applications. At low signal-to-noise ratio (SNR) the deviations from the assumed noise characteristics are critical and the degradation may be severe for the localization result. In this chapter, we present an algorithm to estimate the noise with band covariance matrix. This algorithm is based on the noise subspace spanned by the eigenvectors associated with the smallest eigenvalues of the covariance matrix of the recorded data. The goal of this study is to investigate how perturbations in the assumed noise covariance matrix affect the accuracy of the narrow-band signal DOA estimates (Stoica et al., 1994). A maximum likelihood algorithm is presented in (Wax, 1991), where the spatial noise covariance is modeled as a function of certain unknown parameters. Also in (Ye & DeGroat, 1995) a maximum likelihood estimator is analyzed. The problem of incomplete pre-whitening or colored noise is circumvented by modeling the noise with a simple descriptive model. There are other approaches to the problem of spatially correlated noise: one is based on the assumption that the correlation structure of the noise field is invariant under a rotation or a translation of the array, while another is based on a certain linear transformation of the sensor output vectors (Zhang & Ye, 2008; Tayem et al., 2006). These methods do not require the estimation of the noise correlation function, but they may be quite sensitive to the deviations from the invariance assumption made, and they are not applicable when the signals also satisfy the invariance assumption.

2. Problem formulation

Consider an array of N sensors which receive the signals in one wave field generated by P ($P < N$) sources in the presence of an additive noise. The received signal vector is sampled and the DFT algorithm is used to transform the data into the frequency domain. We represent these samples by:

$$\mathbf{r}(f) = \mathbf{A}(f)\mathbf{s}(f) + \mathbf{n}(f) \tag{1}$$

where $\mathbf{r}(f)$, $\mathbf{s}(f)$ and $\mathbf{n}(f)$ are respectively the Fourier transforms of the array outputs, the source signals and the noise vectors. The $\mathbf{A}(f)$, matrix of dimensions $(N \times P)$ is the transfer matrix of the source-sensor array systems with respect to some chosen reference point. The sensor noise is assumed to be independent of the source signals and partially spatially correlated. The sources are assumed to be uncorrelated. The covariance matrix of the data can be defined by the $(N \times N)$-dimensional matrix.

$$\mathbf{\Gamma}(f) = E[\mathbf{r}(f)\mathbf{r}^{+}(f)] \tag{2}$$

$$\mathbf{\Gamma}(f) = \mathbf{A}(f)\mathbf{\Gamma}_s(f)\mathbf{A}^{+}(f) + \mathbf{\Gamma}_n(f) \tag{3}$$

Where $E[.]$ denotes the expectation operator, superscript $+$ represents conjugate transpose, $\mathbf{\Gamma}_n(f) = E[\mathbf{n}(f)\mathbf{n}^{+}(f)]$ is the $(N \times N)$ noise covariance matrix, and $\mathbf{\Gamma}_s(f) = E[\mathbf{s}(f)\mathbf{s}^{+}(f)]$ is the $(P \times P)$ signal covariance matrix. The above assumption concerning the non-correlation of the sources means that $\mathbf{\Gamma}_s(f)$ is full rank.

The High-Resolution algorithms of array processing assume that the matrix $\mathbf{\Gamma}_n(f)$ is diagonal. The subspace-based techniques are based on these properties. For example the MUSIC (Multiple Signal Classification)(Cadzow, 1998) null-spectrum $P_{music}(\theta)$ is defined by :

$$P_{music}(\theta) = \frac{1}{|\mathbf{a}^{+}(\theta)\hat{\mathbf{V}}_N(f)\hat{\mathbf{V}}_N^{+}(f)\mathbf{a}(\theta)|} \tag{4}$$

and it is expected that $P_{music}(\theta)$ has maximum points around $\theta \in \{\theta_1, ..., \theta_P\}$, where $\theta_1, ..., \theta_p$ are the directions of arrival of the sources and $\hat{\mathbf{V}}_N(f) = \{\mathbf{v}_{P+1}(f) \dots \mathbf{v}_N(f)\}$. Therefore, we can estimate the DOA by taking the local maximum points of $P_{music}(\theta)$.

In this chapter, we consider that the matrix $\mathbf{\Gamma}_n(f)$ is not diagonal because the noise realizations are spatially correlated and then the performances of these methods are considerably degraded.

3. Modeling the noise field

A fundamental limitation of the standard parametric array processing algorithms is that the covariance matrix of background noise cannot, in general, be estimated along with the signal parameters. So this leads to an unidentifiable parametrization, the measured data should always be regarded to consist of only the noise with a covariance matrix equal to that of the observed sample. This is a reason for imposing a model on the background noise. Several parametric noise models have been proposed in some literatures recently. Here, as well as in (Zhang & Ye, 2008), a descriptive model will be used, that is, the spatial noise covariance matrix is assumed to consist of a linear combination of some unknown parameters, which are weighted by known basis matrices. There are two different noise phenomenons to be described. We can model the noise as:

- an internal noise generated by the sensors so-called thermal noise. This noise is assumed to be independent (Zhang & Ye, 2008; Werner & Jansson, 2007) from sensor to sensor, but not necessarily spatially white. Then the spatial covariance matrix of this noise denoted $\mathbf{\Gamma}_n^S(f)$ is diagonal.

- an external noise received on the sensors, whose spatial covariance matrix is assumed to have the following structure (Zhang & Ye, 2008; Werner & Jansson, 2007; Friedlander & Weiss, 1995), $\mathbf{\Gamma}_n^B(f) = \sum_{k=1}^{K} \alpha_k \beta_k$, where α_k are unknown parameters and β_k are complex weighting matrices, β_k are chosen such that $\mathbf{\Gamma}_n^B(f)$ is positive definite and of band structure.

Consequently, the additive noise is the sum of these two noise terms and the spatial covariance matrix is

$$\mathbf{\Gamma}_n(f) = \mathbf{\Gamma}_n^S(f) + \mathbf{\Gamma}_n^B(f) \quad (5)$$

4. Modeling the covariance matrix of the band noise

In many applications when a uniform linear array antenna system is used, it is reasonable to assume that noise correlation is decreasing along the array (see Fig. 1). This is a widely used model for colored noise. We can then obtain a specific model for noise correlation under the following assumptions:

- the correlation length is K which means that the spatial correlation attains up to the $K-$th sensor;
- the noise realizations received by sensors which are separated with a distance no less than Kd, where d is the distance between sensors, are considered uncorrelated,

The noise correlation model which is obtained is represented on Fig. 1.

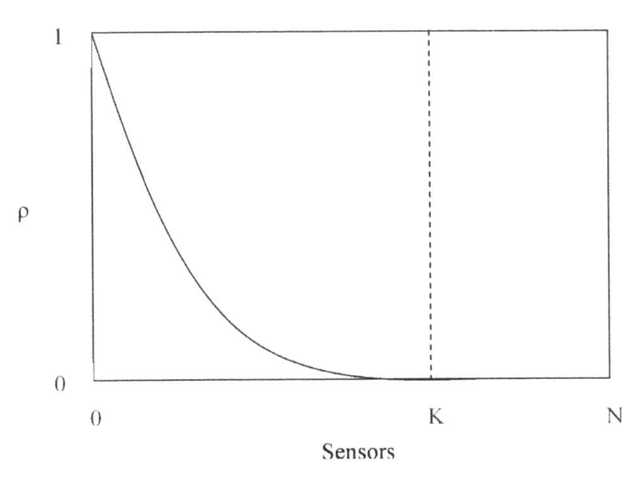

Fig. 1. Noise correlation along an uniform linear array with N sensors, ρ is the noise spatial correlation coefficient.

In this chapter the noise covariance matrix is modeled as an Hermitian, positive-definite band matrix $\Gamma_n(f)$, with half-bandwidth K. The $(i,m)-$th element of $\Gamma_n(f)$ is ρ_{mi} with:

$$\rho_{mi} = 0, \quad for \quad |i-m| \geq K \quad i,m = 1,\ldots,N$$

$$\Gamma_n = \begin{pmatrix} \sigma_1^2(f) & \rho_{12}(f) & \cdots \\ \rho_{12}^*(f) & \sigma_2^2(f) & \cdots \\ \vdots & \ddots & \ddots \\ \rho_{1K}^*(f) & \cdots & \rho_{K-1}^*(f) \\ \vdots & \ddots & \ddots \\ 0 & \cdots & \rho_{KN}^*(f) \\ \rho_{1K}(f) & \cdots & 0 \\ \rho_{1(K+1)}(f) & \cdots & 0 \\ \cdots & \ddots & \vdots \\ \sigma_K^2(f) & \cdots & \rho_{KN}(f) \\ \cdots & \ddots & \vdots \\ \cdots & \rho_{K(N-1)}^*(f) & \sigma_N^2(f) \end{pmatrix}$$

Where $\rho_{mi} = \bar{\rho}_{mi} + j\tilde{\rho}_{mi}$; $i,m=1,...,N$; ρ_{mi} are complex variables, $j^2 = -1$ and σ_i^2 is the noise variance at the ith sensor, and $*$ denotes complex conjugate.
In the following section, an algorithm to estimate the band noise covariance matrix is developed for narrow-band signals.

5. Estimation of the noise covariance matrix

5.1 Proposed algorithm

Several methods have been proposed for estimating the directions of arrival of multiple sources in unknown noise fields. Initially the noise spectral matrix is measured, when signals of interest are not present. Other techniques (Abeidaa & Delmas, 2007) based on the maximum likelihood algorithm are developed, which incorporate a noise model to reduce the bias for estimating both the noise covariance matrix and the directions of arrival of the sources.

Our approach is realized in two steps. Using an iterative algorithm, the noise covariance matrix is estimated, then this estimate is subtracted from the covariance matrix of the received signals.

The proposed algorithm for estimating the noise covariance matrix can be summarized as follows:

Step 1 : Estimate the covariance matrix $\Gamma(f)$ of the received signals by the expectation of T time measures noted by $\hat{\Gamma}(f)$. $\hat{\Gamma}(f) = \frac{1}{T}\left[\sum_{t=1}^{T} \mathbf{r_t(f)r_t^+(f)}\right]$. The eigendecomposition of this matrix is given by:

$\hat{\Gamma}(f) = \mathbf{V}(f)\Lambda(f)\mathbf{V}^+(f)$ with $\Lambda(f) = diag[\lambda_1(f),...,\lambda_N(f)]$ and $\mathbf{V}(f) = [\mathbf{v}_1(f), \mathbf{v}_2(f),..., \mathbf{v}_N(f)]$ where $\lambda_i(f), i = 1...N, (\lambda_1 \geq \lambda_2 \geq,..., \geq \lambda_N)$, and $\mathbf{v}_i(f)$ are respectively the i-th eigenvalue and the corresponding eigenvector.

And we initialize the noise covariance matrix by $\Gamma_n^0(f) = \mathbf{0}$.

Step 2 : Calculate $\mathbf{W}_P = \mathbf{V}_S(f)\Lambda_S^{1/2}(f)$, and $\mathbf{V}_S(f) = [\mathbf{v}_1(f), \mathbf{v}_2(f),..., \mathbf{v}_P(f)]$ is the matrix of the P eigenvectors associated with the first P largest eigenvalues of $\hat{\Gamma}(f)$.
Let $\Delta^1 = \mathbf{W}_P(f)\mathbf{W}_P^+(f)$.

Step 3 : Calculate the (i,j)th element of the current noise covariance matrix

$$[\Gamma_n^1(f)]_{ij} = [\hat{\Gamma}(f) - \Delta^1]_{ij} \;\; if \;\; | \, i-j \, | < K \;\; i,j = 1,...,N$$

and

$$[\Gamma_n^1(f)]_{ij} = 0 \;\; if \;\; | \, i-j \, | \geq K$$

$$\Gamma_n^l(f) = \begin{pmatrix} \Gamma_{11}^l(f) - \Delta_{11}^l(f) & \cdots & \Gamma_{1K}^l(f) - \Delta_{1K}^l(f) \\ \vdots & \ddots & \ddots \\ \Gamma_{K1}^l(f) - \Delta_{K1}^l(f) & \cdots & \Gamma_{KK}^l(f) - \Delta_{KK}^l(f) \\ \vdots & \ddots & \ddots \\ 0 & \cdots & \cdots \\ \cdots & 0 & \\ \cdots & \vdots & \\ \cdots & \Gamma_{NK}^l(f) - \Delta_{NK}^l(f) & \\ \cdots & \vdots & \\ \cdots & \Gamma_{NN}^l(f) - \Delta_{NN}^l(f) & \end{pmatrix}$$

Step 4 : Eigendecomposition of the matrix $[\hat{\Gamma}(f) - \Gamma_n^1(f)]$. The new matrices $\Delta^2(f)$ and $\Gamma_n^2(f)$ are calculated using the previous steps. Repeat the algorithm until a significant improvement of the estimated noise covariance matrix is obtained.

Stop test : The iteration is stopped when $\|\Gamma_n^{l+1}(f) - \Gamma_n^l(f)\|_F < \epsilon$ with ϵ a fixed threshold. The difference between the Frobenius norms of matrices $\Gamma_n^{l+1}(f)$ and $\Gamma_n^l(f)$ is given by :

$$\|\Gamma_n^{l+1}(f) - \Gamma_n^l(f)\|_F = \left[\sum_{i,j=1}^{N} t_{ij}^2(f) \right]^{1/2}$$

where $t_{ij}(f) = [\Gamma_n^{l+1}(f) - \Gamma_n^l(f)]_{ij}$.

5.2 Spatial correlation length

In the previously proposed iterative algorithm, the spatial correlation length of the noise is supposed to be known. In practice, this is aforehand uncertain, therefore the search for a criterion of an estimate of K is necessary. In (Tayem et al., 2006), one algorithm which jointly estimates the number of sources and the spatial correlation length of the noise is presented. We propose to vary the value of K until the stability of the result is reached, that is, until the noise covariance matrix does not vary when K varies. The algorithm incorporating the choice of the correlation length K is presented in Fig. 2. In the stop test, we check whether $\|[\Gamma_n^{K+1}]_1(f) - [\Gamma_n^K]_1(f)\|_F < \epsilon$ or not.

6. Simulation results

In the following simulations, a uniform linear array of $N = 10$ omnidirectional sensors with equal inter-element spacing $d = \frac{c}{4f_o}$ is used, where f_o is the mid-band frequency and c is the velocity of propagation. The number of independent realizations used for estimating the

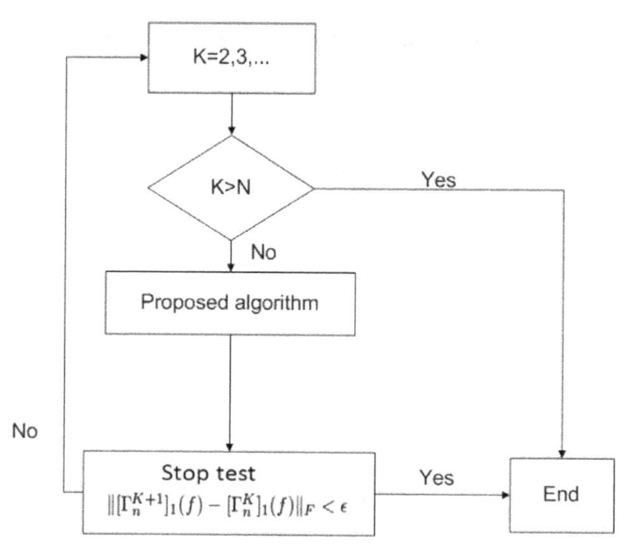

Fig. 2. Integration of the choice of K in the algorithm, where $[\Gamma_n^K]_1(f)$ indicates the principal diagonal of the banded noise covariance matrix $\Gamma_n(f)$ with spatial correlation length K.

covariance matrix of the received signals is 1000. The signal sources are temporally stationary zero-mean white Gaussian processes with the same frequency $f_o = 115\ Hz$. Three equi-power uncorrelated sources impinge on the array from different angles with the $SNR = 10dB$. The noise power is taken as the average of the diagonal elements of the noise covariance matrix $\sigma^2 = \frac{1}{N}\sum_{i=1}^{N}\sigma_i^2$.

To demonstrate the performance of the proposed algorithm, three situations are considered:

- Band-Toeplitz noise covariance matrix, with each element given by a modeling function;
- Band-Toeplitz noise covariance matrix, where the element values are arbitrary;
- Band noise covariance matrix, with each element arbitrary.

In each case, two spatial correlation lengths are studied: $K = 3$ and $K = 5$.

6.1 Noise covariance matrix estimation and results obtained

To localize the directions of arrival of sources and to evaluate the performance of the proposed algorithm, the High-Resolution methods such as MUSIC (Kushki & Plataniotis, 2009; Hawkes & Nehorai, 2001) are used after the preprocessing, with the exact number of sources ($P = 3$). This detection problem is not considered in this study.

Example 1 : *Band-Toeplitz noise covariance matrix:*

In this example, the spatial correlation between the noises is exponentially decreasing along the array of antenna and the elements of the noise covariance matrix are expressed as:

$$[\Gamma_n(f)]_{i,m} = \sigma^2\rho^{|i-m|}e^{j\pi(i-m)/2} \quad if \quad |i-m| < K$$

and,

$$[\Gamma_n(f)]_{i,m} = 0 \qquad if \quad |i-m| \geq K$$

where σ^2 is the noise variance equal for every sensor and ρ is the spatial correlation coefficient. The values which are retained are: $\sigma^2 = 1$ and $\rho = 0.7$.

In each of the two studied cases ($K = 3$ and $K = 5$), the noise covariance matrix is estimated with a fixed threshold value $\epsilon = 10^{-5}$ after a few iterations and we notice that the number of iterations for $K = 5$ is greater than that of $K = 3$.

Example 2 : *Band-Toeplitz noise covariance matrix*

In this example, the covariance matrix elements are chosen such that their values are decreasing along the array of antenna. The noise covariance matrix has the same structure as in example 1:

$$\Gamma_n = \begin{pmatrix} \sigma^2 & \rho_2 & \cdots & \rho_K & \cdots & 0 \\ \rho_2^* & \sigma^2 & \rho_2 & \cdots & \cdots & 0 \\ \vdots & \ddots & \ddots & \ddots & \ddots & \vdots \\ \rho_K^* & \ddots & \ddots & \ddots & \ddots & \vdots \\ \vdots & \ddots & \ddots & \ddots & \ddots & \vdots \\ 0 & \cdots & \rho_K^* & \cdots & \rho_2^* & \sigma^2 \end{pmatrix}$$

The parameters used are: in the case of $K = 3$, $\sigma^2 = 1$, $\rho_2 = 0.4 + 0.3j$ and $\rho_3 = 0.1 + 0.07j$. Using the proposed algorithm, the three complex parameters of the noise covariance matrix can be perfectly estimated.

- For $K = 5$: $\sigma^2 = 1$, $\rho_2 = 0.4 + 0.3j$, $\rho_3 = 0.1 + 0.07j$, $\rho_4 = 0.07 + 0.05j$ and $\rho_5 = 0.01 + 0.009j$. The proposed algorithm gives good estimates of the simulated parameters.

Example 3 : *Band noise covariance matrix with random elements:*

Each element of the band noise covariance matrix is obtained by the average of several simulations simulated with random numbers uniformly distributed in the interval (0,1).

For $K = 3$, Fig. 3 shows the differences between the 10 elements of the principal diagonal of the simulated matrix and those of the estimated matrix.

For $K = 5$, Fig. 4 shows the obtained results. Comparing these two results, we can remark that when K increases the estimation error increases. This concluding remark is observed on many simulations.

Figures 5, 6, 7, 8, 9 and 10 show the localization results of three sources before and after the preprocessing. Before the preprocessing, we use directly the MUSIC method to localize the sources. Once the noise covariance matrix is estimated with the proposed algorithm, this matrix is subtracted from the initial covariance matrix of the received signals, and then we use the MUSIC method to localize the sources. The three simulated sources are 5°, 10° and 20° for Fig. 5 and 6; 5°, 15° and 20° for Figs. 7 and 8; 5°, 15° and 25° for Fig. 9 and 10. For Figs. 7 and 8, the simulated SNR is greater than those of Figs. 5, 6 and Figs. 7 and 8.

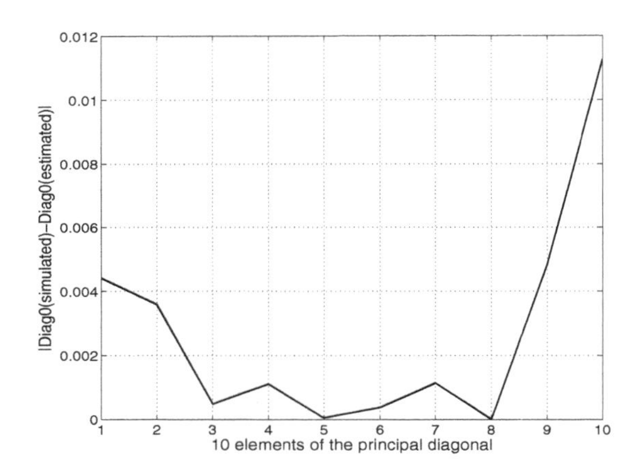

Fig. 3. Variations of the estimation error along the principal diagonal of the noise covariance matrix for $K = 3$.

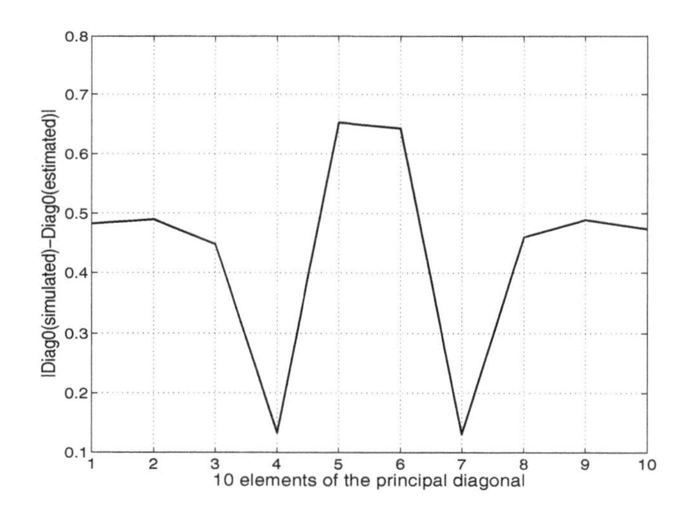

Fig. 4. Variations of the estimation error along the principal diagonal of the noise covariance matrix for $K = 5$.

The comparison of the results of Figs. 5, 6 and 7, 8 comes to the conclusion that the MUSIC method cannot separate the close sources without the preprocessing when the SNR is low, so in Fig. 5 we can only detect two sources before preprocessing. And for each case we can note that there is improvement in the results obtained with the preprocessing. Comparing the results of $K = 3$ with that of $K = 5$ for each figure, we can also reconfirm that when K increases, the estimation error increases on whitening, so we obtain better results with the preprocessing for $K = 3$ than $K = 5$.

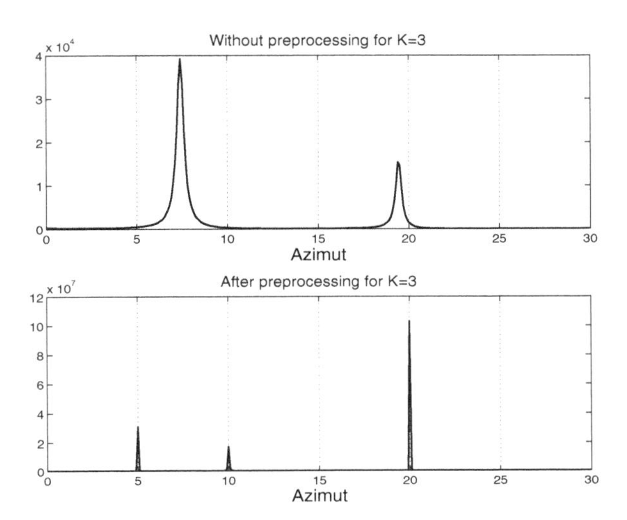

Fig. 5. Localization of the three sources at 5°, 10° and 20° without and with noise pre-processing for $K = 3$.

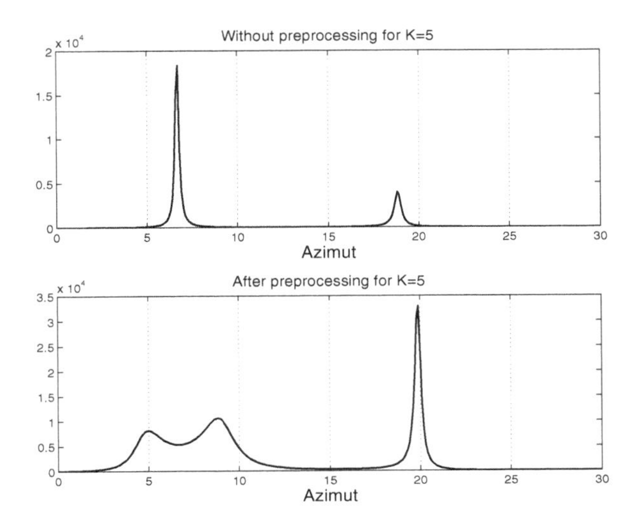

Fig. 6. Localization of the three sources at 5°, 10° and 20° without and with noise pre-processing for $K = 5$.

In order to evaluate the performances of this algorithm, we study, below, the influence of the involved parameters.

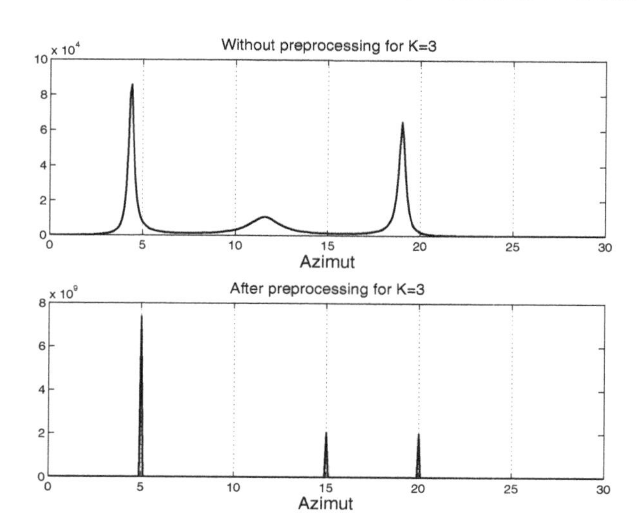

Fig. 7. Localization of the three sources at 5°, 15° and 20° without and with noise pre-processing with greater SNR than figure 5 for $K = 3$.

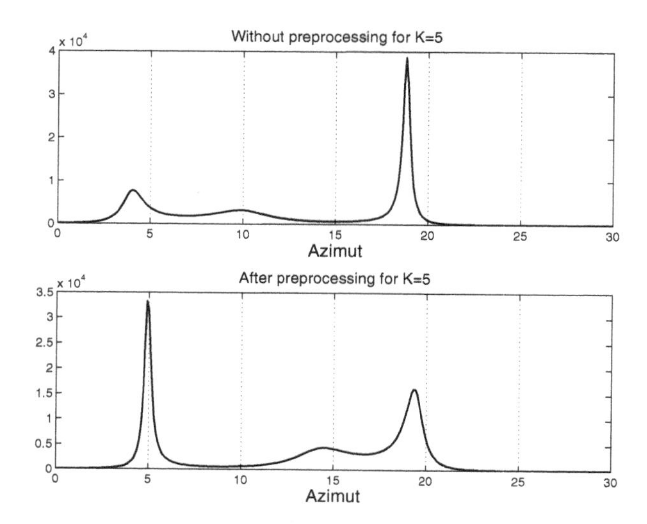

Fig. 8. Localization of the three sources at 5°, 15° and 20° without and with noise pre-processing with greater SNR than figure 6 for $K = 5$.

6.2 Choice of the parameters

6.2.1 Spatial correlation length of the noise

Figure 11 shows the variations of the estimation error of the noise covariance matrix when the spatial correlation length of the noise K is increasing from 2 to N and the number of sources is

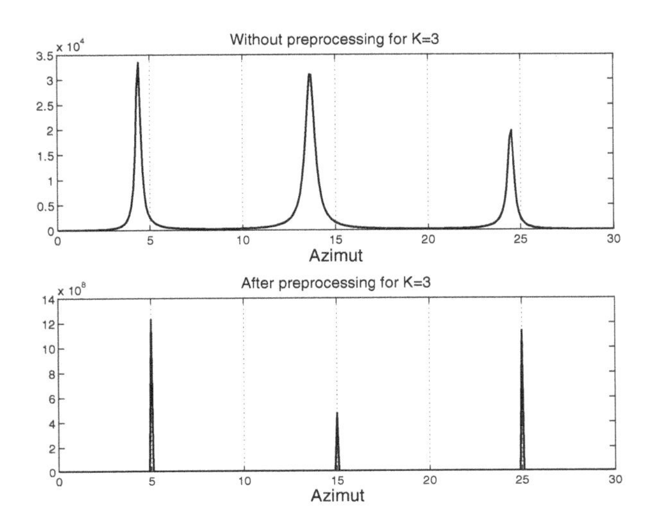

Fig. 9. Localization of the three sources at 5°, 15° and 25° without and with noise pre-processing for $K = 3$.

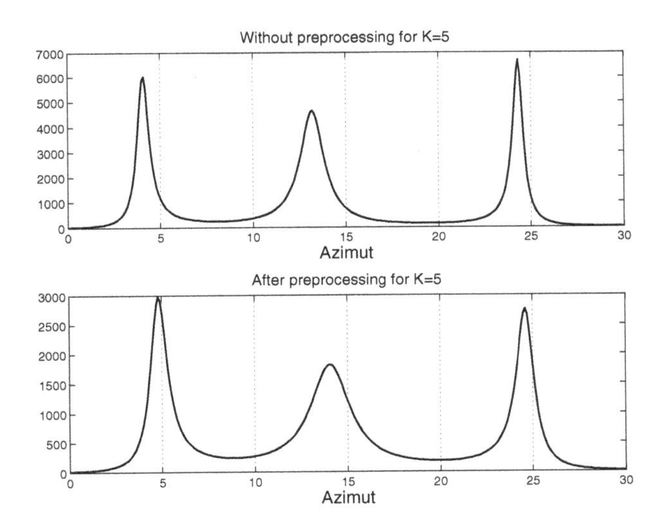

Fig. 10. Localization of the three sources at 5°, 15° and 25° without and with noise pre-processing for $K = 5$.

fixed to be P, $P = 1$ or $P = 9$. This error is defined by: $EE = ||\Gamma_n^{simulated} - \Gamma_n^{estimated}||_F$.

Figure 11 shows that if $P = 1$, the estimation error is null until $K = 5$. On the other hand for $P = 9$, we have an estimation error of the covariance matrix of the noise as soon as $K = 2$ and the error is greater than the error for $P = 1$.

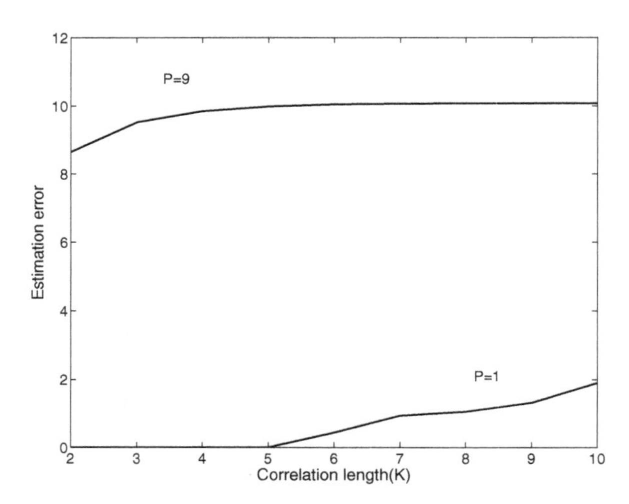

Fig. 11. Estimation error of the covariance matrix of the noise according to its spatial correlation length with P=1 and P=9.

To study the influence of K on the localization, we draw, Fig. 12, according to the spatial correlation length K of the noise, the variations of the bias of estimate of the azimuth in the case of only one source localized at 10°.

We define that the bias of the P estimated directions of the arrival of the sources is calculated by:

$$Bias = \frac{1}{P}\sum_{p=1}^{P} bias(p)$$

where

$$bias(p) = E\left[\theta(p) - \hat{\theta}(p)\right] = \frac{1}{T}\sum_{i=1}^{T} |\theta(i) - \hat{\theta}(i)|$$

The experimental results presented in Figs. 11 and 12 show that the correlation length and the number of sources influence the estimate of the covariance matrix of the noise and then the estimate of the DOA values. We study, below, the influence of the signal-to-noise ratio SNR on the estimate of the covariance matrix of the noise.

6.2.2 Signal-to-noise ratio influence

In order to study the performances of the proposed algorithm according to the signal-to-noise ratio, we plot, Fig. 13, the estimation error over the noise covariance matrix (the estimation error is defined in the section 6.2.1) according to the spatial correlation length K for $SNR = 0\,dB$ and $SNR = 10\,dB$.

We conclude that the choice of the value of K influences the speed and the efficiency of this algorithm. Indeed, many simulations show that this algorithm estimates the matrix quickly, if $K \ll N$. On the other hand if K is close to N, the algorithm requires a great number of iterations. This is due to the increase of the number of unknown factors to estimate. The

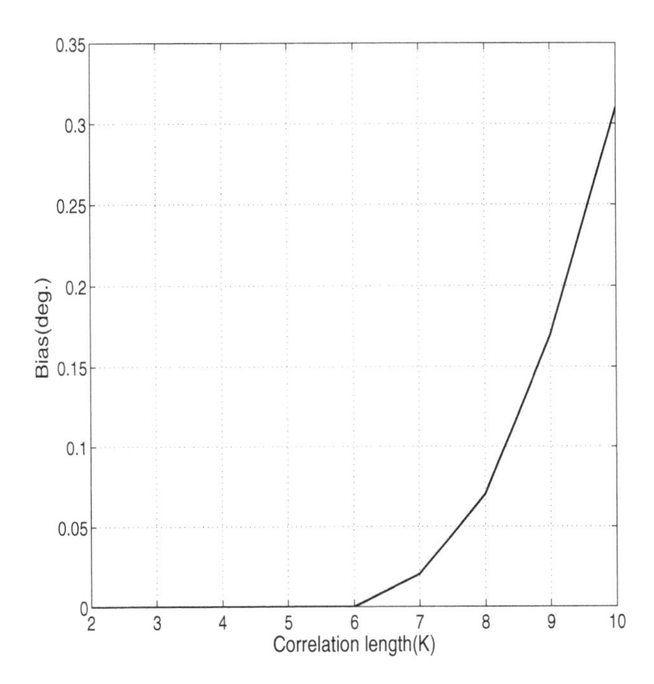

Fig. 12. Bias of the estimated direction according to spatial correlation length of the noise.

efficiency and the speed also depend on the signal-to-noise ratio, the number of sensors, the number of sources to be localized and the angular difference between the sources.

The spatial correlation length K authorized by the algorithm is a function of the number of sensors and the number of sources. Indeed, the number of parameters of the signal to be estimated is P^2 and the number of parameters of the noise is $Nber(K)$. In order to estimate them it is necessary that $N^2 \geq P^2 + Nber(K)$ and that $K \leq N$. In the limit case: $P = N - 1$, we have $Nber(K) \leq 2N - 1$, which corresponds to a bi-diagonal noise covariance matrix. If the model of the noise covariance matrix is band-Toeplitz (Tayem et al., 2006; Werner & Jansson, 2007), the convergence of the proposed algorithm is fast, and the correlation length of noise can reach N.

7. Experimental data

The studied signals are recorded during an underwater acoustic experiment. The experiment is carried out in an acoustic tank under conditions which are similar to those in a marine environment. The bottom of the tank is filled with sand. The experimental device is presented in figure 14. A source emits a narrow-band signal ($f_o = 350\ KHz$). In addition to the signal source a spatially correlated Gaussian noise is emitted. The signal-to-noise ratio is 5 dB. Our objective is to estimate the directions of arrival of the signals during the experiment. The signals are received on one uniform linear array. The observed signals come from various reflections on the objects being in the tank. Generally the aims of acousticians is the detection,

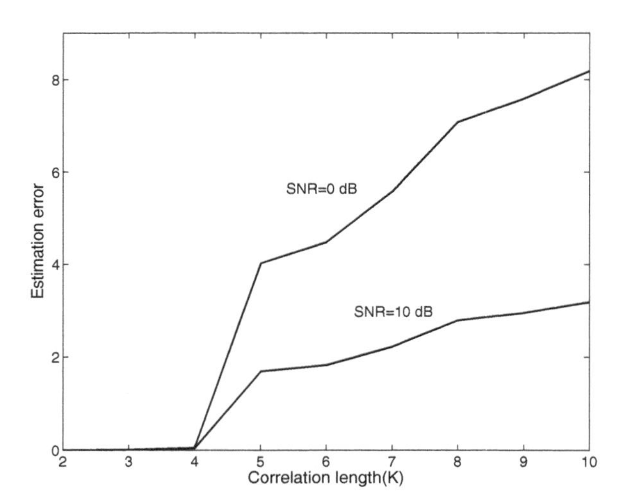

Fig. 13. Estimation error of the covariance matrix of the noise according to the spatial correlation length K and the signal-to-noise ratio SNR.

localization and identification of these objects. In this experiment we have recorded the reflected signals by a single receiver. This receiver is moved along a straight line between position $X_{min} = 50mm$ and position $X_{max} = 150mm$ with a step of $\alpha = 1mm$ in order to create a uniform linear array (see figure 15).

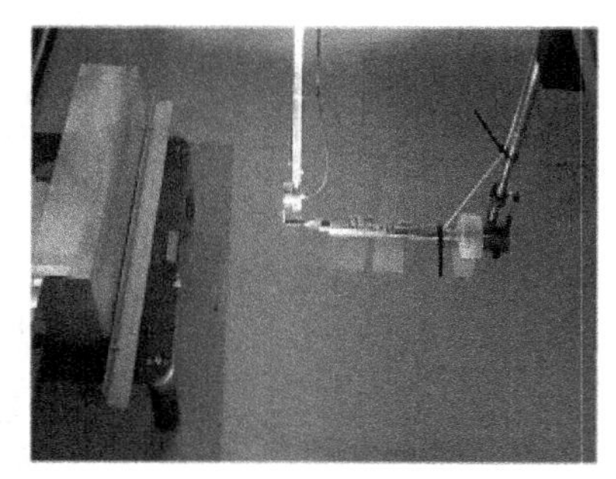

Fig. 14. Experimental device.

Two objects are placed at the bottom of the tank and the emitting source describes a circular motion with a step of 0.5° by covering the angular band going from 1° to 8.5°. The signals received when the angle of emission is $\theta=5^{\circ}$ are shown in figure 16. This figure shows that there exists two paths, which may correspond to the reflected signals on the two objects. The results of the localization are given in figures 17 and 18. We note that in spite of the presence of the correlated noise our algorithm estimate efficiently the DOA of the reflected signals during the experiment.

Figure 17 shows the obtained results of the localization using MUSIC method on the covariance matrices. The DOA of the reflected signals on the two objects are not estimated. This is due to the fact that the noise is correlated.

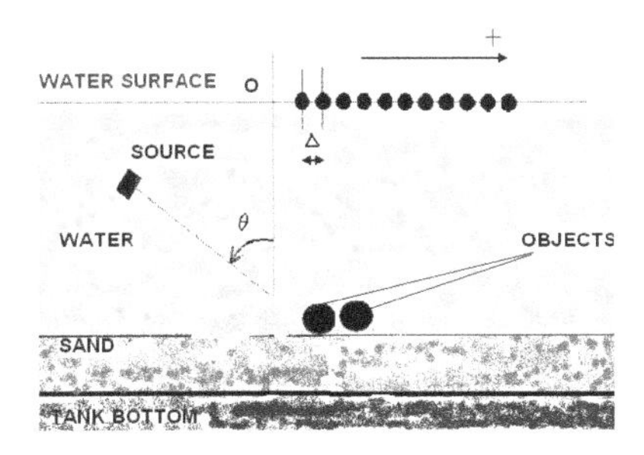

Fig. 15. Experimental setup.

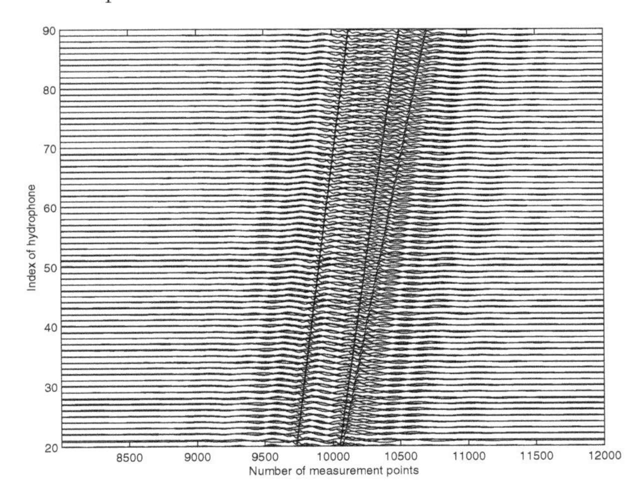

Fig. 16. Received signals.

Figure 18 shows the obtained results using our algorithm. The two objects are localized.

8. Cumulant based coherent signal subspace method for bearing and range estimation in noisy environment

In the rest of this chapter, we consider an array of N sensors which received the signals emitted by P wide band sources ($N > P$) in the presence of an additive Gaussian noise. The received signal vector, in the frequency domain, is given by

$$\mathbf{r}(f_n) = \mathbf{A}(f_n)\mathbf{s}(f_n) + \mathbf{n}(f_n), \quad \text{for } n = 1, ..., M \tag{6}$$

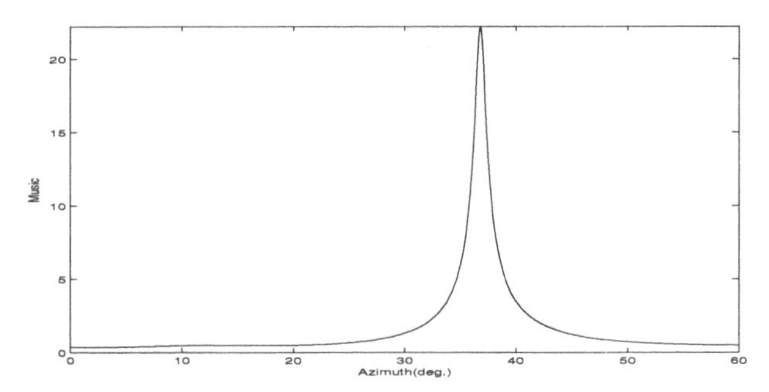

Fig. 17. Localization results with MUSIC.

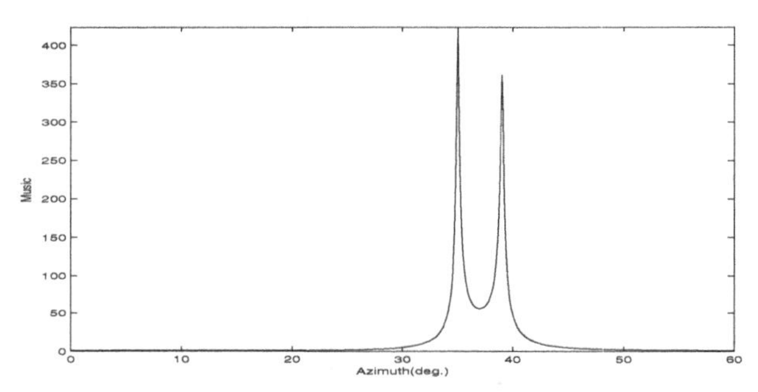

Fig. 18. Localization results with proposed algorithm.

where $\mathbf{A}(f_n) = [\mathbf{a}(f_n, \theta_1), \mathbf{a}(f_n, \theta_2), ..., \mathbf{a}(f_n, \theta_P)]$, $\mathbf{s}(f_n) = [s_1(f_n), s_2(f_n), ..., s_P(f_n)]^T$, and $\mathbf{n}(f_n) = [n_1(f_n), n_2(f_n), ..., n_N(f_n)]^T$.
$\mathbf{r}(f_n)$ is the Fourier transforms of the array output vector, $\mathbf{s}(f_n)$ is the vector of zero-mean complex random non-Gaussian source signals, assumed to be stationary over the observation interval, $\mathbf{n}(f_n)$ is the vector of zero-mean complex white Gaussian noise and statistically independent of signals and $\mathbf{A}(f_n)$ is the transfer matrix (steering matrix) of the source sensor array systems computed by the $\mathbf{a}_k(f_n)$ $(k = 1, ..., P)$ source steering vectors, assumed to have full column rank. In addition to the model (6), we also assume that the signals are statistically independent. In this case, a fourth order cumulant is given by

$$\begin{aligned} \text{Cum}(r_{k_1}, r_{k_2}, r_{l_1}, r_{l_2}) =& \text{E}\{r_{k_1} r_{k_2} r^*_{l_1} r^*_{l_2}\} - \text{E}\{r_{k_1} r^*_{l_1}\}\text{E}\{r_{k_2} r^*_{l_2}\} \\ & - \text{E}\{r_{k_1} r^*_{l_2}\}\text{E}\{r_{k_2} r^*_{l_1}\} \end{aligned} \quad (7)$$

where r_{k_1} is the k_1 element in the vector $\mathbf{r}$ The indices k_2, l_1, l_2 are similarly defined. The cumulant matrix consisting of all possible permutations of the four indices $\{k_1, k_2, l_1, l_2\}$ is given in (Yuen & Friedlander, 1997) as

$$\mathbf{C}_g(f_n) \triangleq \sum_{g=1}^{P} \Big(\mathbf{a}_g(f_n) \otimes \mathbf{a}^*_g(f_n)\Big) u_g(f_n) \Big(\mathbf{a}_g(f_n) \otimes \mathbf{a}^*_g(f_n)\Big)^+ \quad (8)$$

where $u_g(f_n)$ is the source kurtosis (i.e., fourth order analog of variance) defined by $u_g(f_n) = \text{Cum}(s_g(f_n), s^*_g(f_n), s_g(f_n), s^*_g(f_n))$ of the gth complex amplitude source and $\otimes$ is the Kronecker product, Cum(.) denotes the cumulant.

When there are N array sensors, $\mathbf{C}_g(f_n)$ is $(N^2 \times N^2)$ matrix. The rows of $\mathbf{C}_g(f_n)$ are indexed by $(k_1 - 1)N + l_1$, and the columns are indexed by $(l_2 - 1)N + k_2$. In terms of the vector $\mathbf{r}(f_n)$, the cumulant matrix $\mathbf{C}_g(f_n)$ is organized compatibly with the matrix $\mathrm{E}\{(\mathbf{r}(f_n) \otimes \mathbf{r}^*(f_n))(\mathbf{r}(f_n) \otimes \mathbf{r}^*(f_n))^+\}$. In other words, the elements of $\mathbf{C}_g(f_n)$ are given by $\mathbf{C}_g((k_1 - 1)N + l_1, (l_2 - 1)N + k_2)$ for $k_1, k_2, l_1, l_2 = 1, 2, ..., N$ and

$$\mathbf{C}_g((k_1 - 1)N + l_1, (l_2 - 1)N + k_2) = \mathrm{Cum}(r_{k_1}, r_{k_2}, r_{l_1}, r_{l_2}) \tag{9}$$

where r_i is the ith element of the vector $\mathbf{r}$. In order to reduce the calculating time, instead of using of the cumulant matrix $(N^2 \times N^2)$ $\mathbf{C}_g(f_n)$, a cumulant slice matrix $(N \times N)$ of the observation vector at frequency f_n can be calculated and it offers the same algebraic properties of $\mathbf{C}_g(f_n)$. This matrix is denoted $\mathbf{C}_1(f_n)$. If we consider a cumulant slice, for example, by using the first row of $\mathbf{C}_g(f_n)$ and reshape it into an $(N \times N)$ hermitian matrix, i.e.

$$\begin{aligned} \mathbf{C}_1(f_n) &\triangleq \mathrm{Cum}(r_1(f_n), r_1^*(f_n), \mathbf{r}(f_n), \mathbf{r}^+(f_n)) \\ &= \begin{bmatrix} c_{1,1} & c_{1,N+1} & \cdots & c_{1,N^2-N+1} \\ c_{1,2} & c_{1,N+2} & \cdots & c_{1,N^2-N+2} \\ \vdots & \vdots & \vdots & \vdots \\ c_{1,N} & c_{1,2N} & \cdots & c_{1,N^2} \end{bmatrix} \\ &= \mathbf{A}(f_n)\mathbf{U}_s(f_n)\mathbf{A}^+(f_n) \end{aligned} \tag{10}$$

where $c_{1,j}$ is the $(1, j)$ element of the cumulant matrix $\mathbf{C}_g(f_n)$ and $\mathbf{U}_s(f_n)$ is the diagonal kurtosis matrix, its ith element is defined as $\mathrm{Cum}(s_i(f_n), s_i^*(f_n), s_i(f_n), s_i^*(f_n))$ with $i = 1, ..., P$.

$\mathbf{C}_1(f_n)$ can be reported as the classical covariance or spectral matrix of received data

$$\mathbf{\Gamma}_r(f_n) = \mathrm{E}[\mathbf{r}(f_n)\mathbf{r}^+(f_n)] = \mathbf{A}(f_n)\mathbf{\Gamma}_\mathbf{s}(f_n)\mathbf{A}^+(f_n) + \mathbf{\Gamma}_\mathbf{n}(f_n) \tag{11}$$

where $\mathbf{\Gamma}_n(f_n) = \mathrm{E}[\mathbf{n}(f_n)\mathbf{n}^+(f_n)]$ is the spectral matrix of the noise vector and $\mathbf{\Gamma}_s(f_n) = \mathrm{E}[\mathbf{s}(f_n)\mathbf{s}^+(f_n)]$ is the spectral matrix of the complex amplitudes $\mathbf{s}(f_n)$.
If the noise is white then: $\Gamma_n(f_n) = \sigma_n^2(f_n)\mathbf{I}$, where $\sigma_n^2(f_n)$ is the noise power and $\mathbf{I}$ is the $(N \times N)$ identity matrix. The signal subspace is shown to be spanned by the P eigenvectors corresponding to P largest eigenvalues of the data spectral matrix $\Gamma_r(f_n)$. But in practice, the noise is not often white or its spatial structure is unknown, hence the interest of the high order statistics as shown in equation (10) in which the fourth order cumulants are not affected by additive Gaussian noise (i.e., $\Gamma_n(f_n) = \mathbf{0}$), so as no noise spatial structure assumption is necessary. If the eigenvalues and the corresponding eigenvectors of $\mathbf{C_1}(\mathbf{f_n})$ are denoted by $\{\lambda_i(f_n)\}_{i=1..N}$ and $\{\mathbf{v_i}(\mathbf{f_n})\}_{\mathbf{i=1..N}}$. Then, the eigendecomposition of the cumulant matrix $\mathbf{C_1}(\mathbf{f_n})$ is exploited so as

$$\mathbf{C_1}(\mathbf{f_n}) = \sum_{i=1}^{N} \lambda_i(f_n)\mathbf{v_i}(\mathbf{f_n})\mathbf{v_i}^+(\mathbf{f_n}) \tag{12}$$

In matrix representation, equation (12) can be written

$$\mathbf{C}_1(f_n) = \mathbf{V}(f_n)\mathbf{\Lambda}(f_n)\mathbf{V}^+(f_n) \tag{13}$$

where $\mathbf{V}(f_n) = [\mathbf{v}_1(f_n), ..., \mathbf{v}_N(f_n)]$ and $\mathbf{\Lambda}(f_n) = diag(\lambda_1(f_n), ..., \lambda_N(f_n))$.

Assuming that the columns of $\mathbf{A}(f_n)$ are all different and linearly independent it follows that for nonsingular $\mathbf{C}_1(f_n)$, the rank of $\mathbf{A}(f_n)\mathbf{U}_s(f_n)\mathbf{A}^+(f_n)$ is P. This rank property implies that:

- the $(N-P)$ multiplicity of its smallest eigenvalues : $\lambda_{P+1}(f_n) = \ldots = \lambda_N(f_n) \cong 0$.

- the eigenvectors corresponding to the minimal eigenvalues are orthogonal to the columns of the matrix $\mathbf{A}(f_n)$, namely, the steering vectors of the signals

$$\mathbf{V}_n(f_n) \triangleq \{\mathbf{v}_{P+1}(f_n) \ldots \mathbf{v}_N(f_n)\} \perp \{\mathbf{a}(\theta_1, f_n) \ldots \mathbf{a}(\theta_P, f_n)\} \tag{14}$$

The eigenstructure based techniques are based on the exploitation of these properties. Then the directions of arrival of the sources are obtained, at the frequency f_n, by the peak positions in a so-called spatial spectrum (MUSIC)

$$\mathbf{Z}(f_n, \theta) = \frac{1}{\mathbf{a}^+(f_n, \theta)\mathbf{V}_n(f_n)\mathbf{V}_n^+(f_n)\mathbf{a}(f_n, \theta)} \tag{15}$$

For locating the wide band sources several solutions have been proposed in literature and are regrouped in this chapter into two groups:

- the incoherent subspace method: the analysis band is divided into several frequency bins and then at each frequency any narrow-band source localization algorithm can be applied and the obtained results are combined to obtain the final result. In addition of a significant calculating time, these methods do not carry out an actual wide band processing, but integrating information as well as possible coming from the various frequencies available to carry out the localization. It follows, for example, that the treatment gain does not increase with the analysis bandwidth. Indeed, it depends only on the resolving power narrow band processing for each considered frequency bin separately

- the coherent subspace method: the different subspaces are transformed in a predefined subspace using the focusing matrices (Valaee & Kabal, 1995; Hung & Kaveh, 1988). Then the obtained subspace is used to estimate the source parameters.

In the following section, the coherent subspace methods are described.

9. Coherent subspace methods

In the high resolution algorithm, the signal subspace is defined as the column span of the steering matrix $\mathbf{A}(f_n)$ which is function of the frequency f_n and the angles-of-arrival. Thus, the signal subspaces at different frequency bins are different. The coherent subspace methods (Hung & Kaveh, 1988) combine the different subspaces in the analysis band by the use of the focusing matrices. The focusing matrices $\mathbf{T}(\mathbf{f_o}, \mathbf{f_n})$ compensate the variations of the transfer matrix with the frequency. So these matrices verify

$$\mathbf{T}(f_o, f_n)\mathbf{A}(f_n) = \mathbf{A}(f_o), \quad for \quad n = 1, \ldots, M \tag{16}$$

where f_o is the focusing frequency.

Initially, Hung et al. (Hung & Kaveh, 1988) have developed the solutions for equation (16), the proposed solution under constraint $\mathbf{T}(f_o, f_n)\mathbf{T}^+(f_o, f_n) = \mathbf{I}$, is

$$\hat{\mathbf{T}}(f_o, f_n) = \mathbf{V}_l(f_o, f_n)\mathbf{V}_r^+(f_o, f_n) \tag{17}$$

where the columns of $\mathbf{V}_l(f_o, f_n)$ and of $\mathbf{V}_r(f_o, f_n)$ are the left and right singular vectors of the matrix $\mathbf{A}(f_n, \theta_i)\mathbf{A}^+(f_o, \theta_i)$ where θ_i is an initial vector of the estimates of the angles-of-arrival, given by an ordinary beamforming preprocess.

It has been shown that the performances of these methods depend on θ_i (Bourennane et al., 1997). In practice, it is very difficult to obtain the accurate estimate of the DOAs. So in order to resolve this initialization problem, the Two-Sided Correlation Transformation (TCT) algorithm is proposed in (Valaee & Kabal, 1995). The focusing matrices $\mathbf{T}(f_o, f_n)$ are obtained by minimizing

$$\begin{cases} \min\limits_{\mathbf{T}(f_o,f_n)} \|\mathbf{P}(f_o) - \mathbf{T}(f_o, f_n)\mathbf{P}(f_n)\mathbf{T}^+(f_o, f_n)\|_F \\ \qquad s.t \quad \mathbf{T}^+(f_o, f_n)\mathbf{T}(f_o, f_n) = \mathbf{I} \end{cases}$$

where $\mathbf{P}(.)$ is the array spectral matrix in noise free environment, $\mathbf{P}(.) = \mathbf{A}(.)\Gamma_s(.)\mathbf{A}^+(.)$ and $\|.\|_F$ is the Frobenius matrix norm. The solution (Valaee & Kabal, 1995)is

$$\mathbf{T}(f_o, f_n) = \mathbf{V}(f_o)\mathbf{V}^+(f_n) \tag{18}$$

where $\mathbf{V}(f_o)$ and $\mathbf{V}(f_n)$ are the eigenvector matrices of $\mathbf{P}(f_o)$ and $\mathbf{P}(f_n)$, respectively.

In order to reduce the calculating time for constructing this operator equation (18), an improved solution is developed in (Bendjama et al., 1998), where only the signal subspace is used, however

$$\mathbf{T}(f_o, f_n)\mathbf{V}_s(f_n) = \mathbf{V}_s^+(f_o) \tag{19}$$

so the operator becomes

$$\mathbf{T}(f_o, f_n) = \mathbf{V}_s(f_o)\mathbf{V}_s^+(f_n) \tag{20}$$

where $\mathbf{V}_s(f_n) = [\mathbf{v}_1(f_n), \mathbf{v}_2(f_n), ..., \mathbf{v}_P(f_n)]$ are the eigenvectors corresponding to P largest eigenvalues of the spectral matrix $\mathbf{\Gamma}_r(f_n)$.

Once $\mathbf{\Gamma}_r(f_n)$ and $\mathbf{T}(f_o, f_n), n = 1..., M$ are formed, the estimate of $\hat{\mathbf{\Gamma}}_r(f_o)$ can be written

$$\hat{\mathbf{\Gamma}}_r(f_o) = \frac{1}{M}\sum_{n=1}^{M} \mathbf{T}(f_o, f_n)\mathbf{\Gamma}_r(f_n)\mathbf{T}^+(f_o, f_n) \tag{21}$$

In particular case, when equation (20) is used

$$\begin{aligned} \hat{\mathbf{\Gamma}}_r(f_o) &= \frac{1}{M}\sum_{n=1}^{M} \mathbf{V}_s(f_o)\mathbf{\Lambda}_s(f_n)\mathbf{V}_s^+(f_o) \\ &= \mathbf{V}_s(f_o)\hat{\mathbf{\Lambda}}_s(f_o)\mathbf{V}_s^+(f_o) \end{aligned} \tag{22}$$

where $\hat{\mathbf{\Lambda}}_s(f_o) = \frac{1}{M}\sum_{n=1}^{M} \mathbf{\Lambda}_s(f_n)$ is the arithmetic mean of the largest eigenvalues of the spectral matrices $\mathbf{\Gamma}_r(f_n)$ $(n = 1, ..., M)$.

Note that, the number of sources, P, can be computed by the number of non-zero eigenvalues of $\hat{\mathbf{\Gamma}}_r(f_o)$.

The efficiency of the different focusing algorithms is depended on the prior knowledge of the noise. All the transform matrices solution of equation system (16) are obtained in the presence of white noise or with the known spatial correlation structure.

In practice, as in underwater acoustic this assumption is not fulfilled then the performances of the subspace algorithms are degraded. To improve these methods in noisy data cases, in this paper, an algorithm is proposed to remove the noises. The basic idea is based on the combination of the high order statistics of received data, the multidimensional filtering and the frequential smoothing for eliminating the noise contributions.

10. Cumulant based coherent signal subspace

The high order statistics of received data are used to eliminate the Gaussian noise. For this the cumulant slice matrix expression (10) is computed. Then the $(P \times N)$ matrix denoted $\mathbf{H}(f_n)$ is formed in order to transform the received data with an aim of obtaining a perfect orthogonalization and eliminate the orthogonal noise component.

$$\mathbf{H}(f_n) = \mathbf{\Lambda}_s(f_n)^{-1/2}\mathbf{V}_s^+(f_n) \tag{23}$$

where $\mathbf{V}_s(f_n)$ and $\mathbf{\Lambda}_s(f_n)$ are the P largest eigenvectors and the corresponding eigenvalues of the slice cumulant matrix $\mathbf{C}_1(f_n)$ respectively. We note that it is necessary that the eigenvalues in $\mathbf{\Lambda}_s(f_n)$ be distinct. This is the case when the source kurtosis are different. If they are not, then the proposed algorithm will not provide correct estimates of the subspace signal sources. The $(P \times 1)$ vector of the transformed received data is

$$\mathbf{r}_t(f_n) = \mathbf{H}(f_n)\mathbf{r}(f_n) = \mathbf{H}(f_n)\mathbf{A}(f_n)\mathbf{s}(f_n) + \mathbf{H}(f_n)\mathbf{n}(f_n) \tag{24}$$

Afterwards, the corresponding $P^2 \times P^2$ cumulant matrix can be expressed as

$$\begin{aligned}\mathbf{C}_t(f_n) =& \mathrm{Cum}\left(\mathbf{r}_t(f_n), \mathbf{r}_t^+(f_n), \mathbf{r}_t(f_n), \mathbf{r}_t^+(f_n)\right)\\ =& \Big((\mathbf{HA})(f_n) \otimes (\mathbf{HA})^*(f_n)\Big)\mathbf{U}_s(f_n)\Big((\mathbf{HA})(f_n) \otimes (\mathbf{HA})^*(f_n)\Big)^+\end{aligned} \tag{25}$$

Or

$$\mathbf{C}_t(f_n) = \underbrace{\Big(\mathbf{B}(f_n) \otimes \mathbf{B}^*(f_n)\Big)}_{=\mathbf{B}_\otimes(f_n)}\mathbf{U}_s(f_n)\Big(\mathbf{B}(f_n) \otimes \mathbf{B}^*(f_n)\Big)^+ \tag{26}$$

Using the property (Mendel, 1991)

$$\mathbf{WX} \otimes \mathbf{YZ} = (\mathbf{W} \otimes \mathbf{Y})(\mathbf{X} \otimes \mathbf{Z}) \tag{27}$$

We can show that

$$\begin{aligned}\mathbf{C}_t(f_n) =& \Big(\mathbf{H}(f_n) \otimes \mathbf{H}^*(f_n)\Big)\Big(\mathbf{A}(f_n) \otimes \mathbf{A}^*(f_n)\Big)\mathbf{U}_s(f_n)\\ & \Big(\mathbf{A}(f_n) \otimes \mathbf{A}^*(f_n)\Big)^+\Big(\mathbf{H}(f_n) \otimes \mathbf{H}^*(f_n)\Big)^+\\ =& \Big(\mathbf{H}(f_n) \otimes \mathbf{H}^*(f_n)\Big)\mathbf{A}_\otimes(f_n)\mathbf{U}_s(f_n)\mathbf{A}_\otimes^+(f_n)\Big(\mathbf{H}(f_n) \otimes \mathbf{H}^*(f_n)\Big)^+\end{aligned} \tag{28}$$

Using (26) and (28), we have

$$\mathbf{B}_\otimes(f_n) = \Big(\mathbf{H}(f_n) \otimes \mathbf{H}^*(f_n)\Big)\mathbf{A}_\otimes(f_n) \tag{29}$$

Using the eigenvectors focusing operator defined as

$$\mathbf{T}(f_o, f_n) = \mathbf{V}_{ts}(f_o)\mathbf{V}_{ts}^{+}(f_n) \tag{30}$$

where $\mathbf{V}_{ts}(.)$ are the eigenvectors of the largest eigenvalues of the cumulant matrix $\mathbf{C}_t(.)$. The average cumulant matrix is

$$\begin{aligned}\hat{\mathbf{C}}_t(f_o) &= \frac{1}{M}\sum_{n=1}^{M}\mathbf{T}(f_o, f_n)\mathbf{C}_t(f_n)\mathbf{T}^{+}(f_o, f_n)\\ &= \mathbf{V}_{ts}(f_o)\hat{\mathbf{\Lambda}}_{ts}(f_o)\mathbf{V}_{ts}^{+}(f_o)\end{aligned} \tag{31}$$

where $\hat{\mathbf{\Lambda}}_{ts}(f_o) = \frac{1}{M}\sum_{n=1}^{M}\mathbf{\Lambda}_{ts}(f_n)$ is the arithmetic mean of the first largest eigenvalues of the cumulant matrix $\mathbf{C}_t(f_n)$. It is easy to show that

$$\hat{\mathbf{C}}_t(f_o) = \mathbf{V}_{ts}(f_o)\hat{\mathbf{\Lambda}}_{ts}(f_o)\mathbf{V}_{ts}^{+}(f_o) = \mathbf{B}_{\otimes}(f_o)\hat{\mathbf{U}}_s(f_o)\mathbf{B}_{\otimes}^{+}(f_o) \tag{32}$$

where $\mathbf{B}_{\otimes}(f_o) \triangleq \mathbf{T}(f_o, f_n)\mathbf{B}_{\otimes}(f_n)$ and $\hat{\mathbf{U}}_s(f_o) = \frac{1}{M}\sum_{n=1}^{M}\mathbf{U}_s(f_n)$.

Multiplying both sides by $\mathbf{B}_{\otimes}^{+}(f_o)$, we get

$$\mathbf{B}_{\otimes}^{+}(f_o)\hat{\mathbf{C}}_t(f_o) = \mathbf{B}_{\otimes}^{+}(f_o)\mathbf{B}_{\otimes}(f_o)\hat{\mathbf{U}}_s(f_o)\mathbf{B}_{\otimes}^{+}(f_o) \tag{33}$$

Because the columns of $\mathbf{B}_{\otimes}(f_o)$ are orthogonal and the sources are decorrelated, $\mathbf{B}_{\otimes}^{+}(f_o)\mathbf{B}_{\otimes}(f_o)\hat{\mathbf{U}}_s(f_o)$ is a diagonal matrix which we will denote by $\mathbf{D}(f_o)$, so that we have

$$\mathbf{B}_{\otimes}^{+}(f_o)\hat{\mathbf{C}}_t(f_o) = \mathbf{D}(f_o)\mathbf{B}_{\otimes}^{+}(f_o) \tag{34}$$

Or

$$\hat{\mathbf{C}}_t(f_o)\mathbf{B}_{\otimes}^{+}(f_o) = \mathbf{D}(f_o)\mathbf{B}_{\otimes}(f_o) \tag{35}$$

This equation tells us that the columns of $\mathbf{B}_{\otimes}(f_o)$ are the left eigenvectors of $\hat{\mathbf{C}}_t(f_o)$ corresponding to the eigenvalues on the diagonal of the matrix $\mathbf{D}(f_o)$: however, since $\hat{\mathbf{C}}_t(f_o)$ is Hermitian, they are also the (right) eigenvectors of $\hat{\mathbf{C}}_t(f_o)$. Furthermore, the columns of $\mathbf{V}_{ts}(f_o)$ are also eigenvectors of $\hat{\mathbf{C}}_t(f_o)$ corresponding to the same (non-zero) eigenvectors of the diagonal matrix $\hat{\mathbf{U}}_s(f_o)$. Given that the eigenvalues of $\hat{\mathbf{C}}_t(f_o)$ are different, the orthonormal eigenvectors are unique up to phase term, this will likely be the case if the source kurtosis are different. Hence the difference between $\mathbf{V}_{ts}(f_o)$ and $\mathbf{B}_{\otimes}(f_o)$ is that the columns may be reordered and each column is multiplied by a complex constant.

The information we want is $\mathbf{A}_{\otimes}(f_o)$, which is given by (29)

$$\mathbf{A}_{\otimes}(f_o) = \Big(\mathbf{H}(f_o) \otimes \mathbf{H}^{*}(f_o)\Big)^{\dagger}\mathbf{B}_{\otimes}(f_o) \tag{36}$$

with † denoting the pseudo-inverse of matrix. We do not have the matrix $\mathbf{B}_{\otimes}(f_o)$, but we have the matrix $\mathbf{V}_{ts}(f_o)$. Hence we can obtain a matrix $\mathbf{A}^{'}_{\otimes}(f_o)$ such that

$$\mathbf{A}^{'}_{\otimes}(f_o) = \left(\mathbf{H}(f_o) \otimes \mathbf{H}^{*}(f_o)\right)^{\dagger} \mathbf{V}_{ts}(f_o) \tag{37}$$

Furthermore, we obtain $\hat{\mathbf{A}}(f_o)$ by extracting out the first N rows and the first P columns of $\mathbf{A}^{'}_{\otimes}(f_o)$. The estimate $\hat{\mathbf{A}}(f_o)$ will be permuted and scaled (column-wise) version of $\mathbf{A}(f_o)$.

This algorithm leads to the estimation of the transfer matrix without prior knowledge of the steering vector or the propagation model such as in the classical techniques.

Therefore, the present algorithm for locating the wide band acoustic sources in the presence of unknown noise can be formulated as the following sequence of steps:

1) Form $\mathbf{C}_1(f_n)$ equation (10) and perform its eigendecomposition;
2) Form $\mathbf{H}(f_n)$ equation (23) $n = 1, ..., M$;
3) Calculate $\mathbf{r}_t(f_n)$ equation (24);
4) Form $\mathbf{C}_t(f_n)$ equation (25) and perform its eigendecomposition;
5) Form $\mathbf{T}(f_o, f_n)$ equation (30) $n = 1, ..., M$;
6) Form $\hat{\mathbf{C}}_t(f_o)$ equation (31) and perform its eigendecomposition;
7) Determine $\mathbf{A}^{'}_{\otimes}(f_o)$ equation (37) and $\hat{\mathbf{A}}(f_o)$.

Naturally one can use the high resolution algorithm to estimate the azimuths of the sources. The orthogonal projector $\mathbf{\Pi}(\mathbf{f_o})$ is

$$\mathbf{\Pi}(f_o) = \mathbf{I} - [\hat{\mathbf{A}}^{+}(f_o)\hat{\mathbf{A}}(f_o)]^{-1}\hat{\mathbf{A}}^{+}(f_o) \tag{38}$$

where $\mathbf{I}$ is the (N×N) identity matrix. Hence, the narrow band Music can be directly applied to estimate the DOA of wide band sources according to

$$Z(f_o, \theta) = \frac{1}{\mathbf{a}^{+}(f_o, \theta)\mathbf{\Pi}(f_o)\mathbf{a}(f_o, \theta)} \tag{39}$$

with $\theta \in [-\pi/2, \pi/2]$.

11. Numerical results

In the following simulations, a linear antenna of $N = 10$ equispaced sensors with equal interelement spacing $d = \frac{c}{2f_o}$ is used, where f_o is the mid band frequency and c is the velocity of propagation. Eight ($P = 8$) wide band source signals arrive at directions : $\theta_1 = 2°, \theta_2 = 5°$, $\theta_3 = 10°$, $\theta_4 = 12°$, $\theta_5 = 30°$, $\theta_6 = 32°$ $\theta_7 = 40°$ and $\theta_8 = 42°$, are temporally stationary zero-mean band pass with the same central frequency $f_o = 110Hz$ and the same bandwidth $B = 40Hz$. The additive noise is uncorrelated with the signals. The Signal-to-Noise Ratio (SNR) is defined here as the ratio of the power of each source signal to the average power of the noise variances, equals on all examples to $SNR = 10dB$.

Experiment 1: Improvement of source localization

In order to point out the improvement of the localization of the sources based on the higher order statistics, our first experiment is carried out in the context of the presence of Gaussian noise. Figures 19 and 20 show the obtained localization results of the eight simulated sources. The number of the sources is taken equal to 8. We can conclude that the fourth order statistics of the received data have improved the spatial resolution. Indeed, the eight sources are perfectly separated. This improvement is due to the fact that the noise is removed using the cumulants.

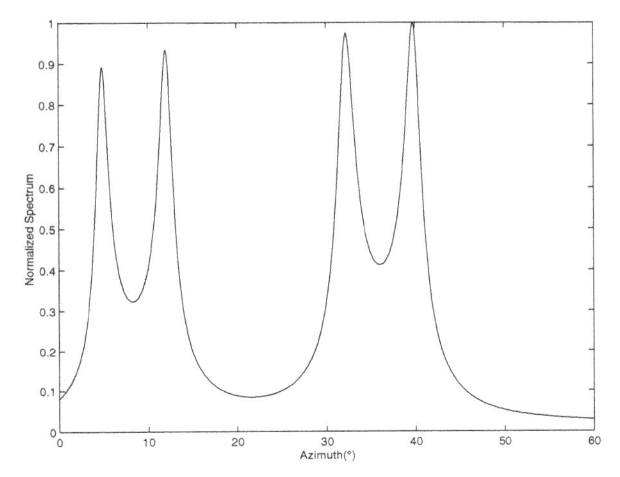

Fig. 19. Spectral matrix-Incoherent signal subspace, 8 uncorrelated sources.

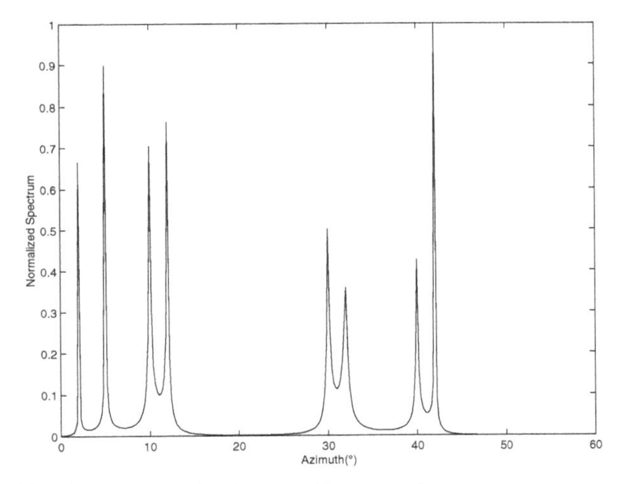

Fig. 20. Cumulant-Incoherent signal subspace, 8 uncorrelated sources.

Experiment 2: Localization of the correlated sources

The localization of a correlated sources is the delicate problem. The spatial smoothing is a solution (Pillai & Kwon, 1989), for the narrow band sources but limited to a linear antenna. In presence of white noise, it is well known that the frequential smoothing leads to localize the wide band correlated sources (Valaee & Kabal, 1995). In this experiment, the eight sources form two fully correlated groups in the presence of Gaussian noise with an unknown spectral matrix. Figures 21 and 22 give the source localization results, the obtained results show that even the coherent signal subspace is used the performance of the high resolution algorithm is degraded (figure 21). Figure 22 shows the cumulant matrix improves the localization, this effectiveness is due to the fact that the noise contribution is null. It follows that the SNR after focusing is improved.

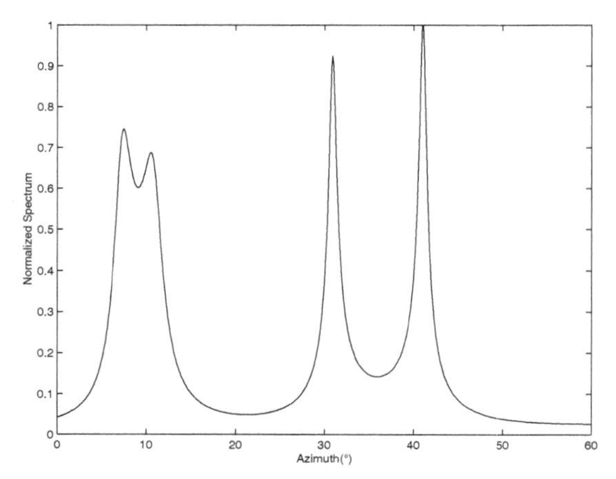

Fig. 21. Spectral matrix-Coherent signal subspace, two groups of correlated sources and Gaussian noise

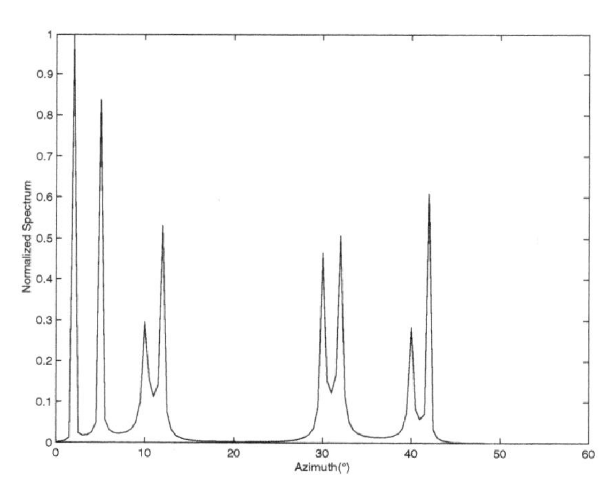

Fig. 22. Cumulant matrix-Coherent signal subspace, two groups of correlated sources and Gaussian noise

Experiment 3: Noise reduction- signal subspace projection

In this part, our algorithm is applied. The noise is modeled as the sum of Gaussian noise and spatially correlated noise. The eight sources are uncorrelated. Figure 23 shows that the cumulant matrix alone is not sufficient to localize the sources. But if the preprocessing of the received data is carried out by the projection of the data on the signal subspace to eliminate the components of the noise which is orthogonal to the signal sources, the DOA will be estimated as shown in figure 24.

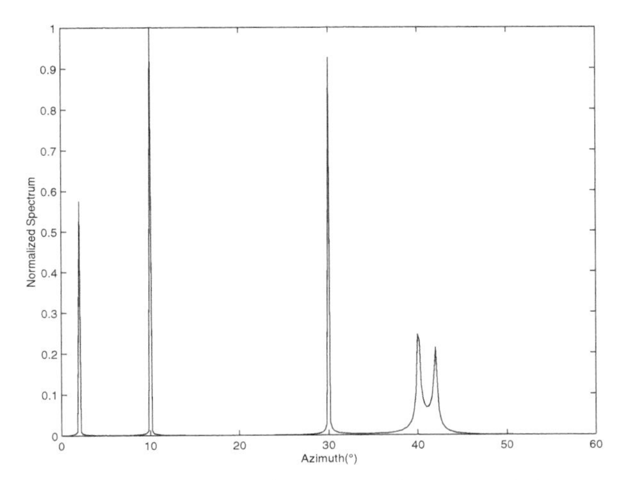

Fig. 23. Cumulant matrix incoherent signal subspace, 8 uncorrelated sources without whitening data

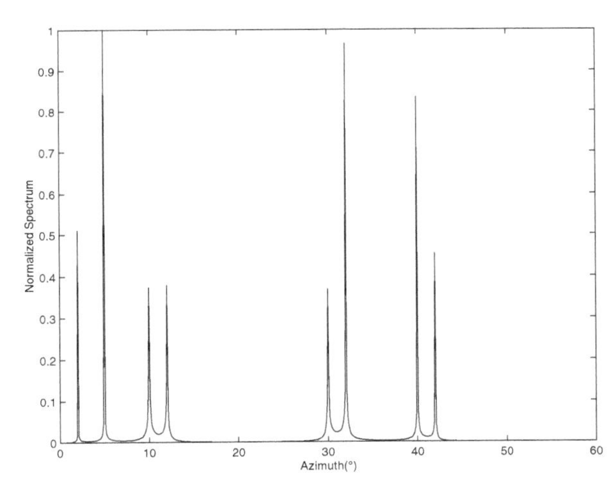

Fig. 24. Cumulant matrix incoherent signal subspace, 8 uncorrelated sources with whitening data

Figure 25 gives the obtained results with the coherent signal subspace when the sources are fully correlated. This last part points out the performance of our algorithm to localize the wide band correlated sources in the presence of unknown noise fields. Note that this results can be considered as an outstanding contribution of our study for locating the acoustic sources. Indeed, our algorithm allows to solve several practical problems. *Proposed algorithm performance:*

In order to study and to compare the performance of our algorithm to the classical wide band methods. For doing so, an experiment is carried out with the same former signals. Figure 26

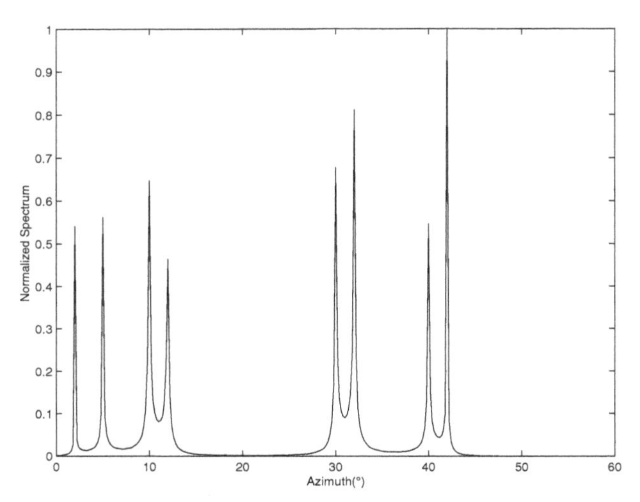

Fig. 25. Coherent cumulant matrices for two groups correlated sources after whitening data

shows the variation of the standard deviation (std) as function of SNR. The std is defined as $std = (1/k)[\sum_{i=1}^{k} |\hat{\theta}_i - \theta_i|^2]^{1/2}$, k is number of independent trials. The SNR is varied from $-40dB$ to $+40dB$ with $k = 500$ independent trials. One can remark the interest of the use of cumulant matrix instead of the spectral matrix and the improvement due to the whitening preprocessing or multidimensional filtering included in our algorithm. Our method after whitening presents the smallest std in all cases.

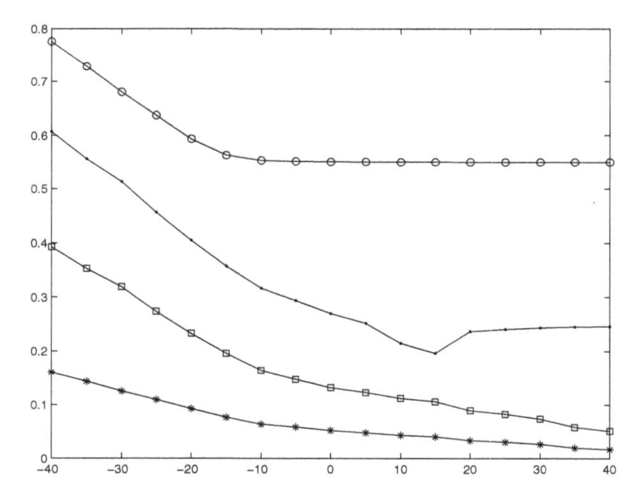

Fig. 26. Standard deviation comparison estimation: $-\circ$ Wang algorithm;

$-.$ TCT algorithm; $-\square$ proposed method without whitening; $-*$ proposed method after whitening.

12. Conclusion

In this chapter the problem of estimating the direction-of-arrival (DOA) of the sources in the presence of spatially correlated noise is studied. The spatial covariance matrix of the noise is modeled as a band matrix and is supposed to have a certain structure. In the numerical example, the noise covariance matrix is supposed to be the same for all sources, which covers many practical cases where the sources are enclosed. This algorithm can be applied to the localization of the sources when the spatial-spectrum of the noise or the spatial correlation function between sensors is known. The obtained results show that the proposed algorithm improves the direction estimates compared to those given by the MUSIC algorithm without preprocessing. Several applications on synthetic data and experiment have been presented to show the limits of these estimators according to the signal-to-noise ratio, the spatial correlation length of the noise, the number of sources and the number of sensors of the array. The motivation of this work is to reduce the computational loads and to reduce the effect of the additive spatially correlated gaussian noise for estimating the DOA of the sources. We also presented methods to estimate the DOA algorithm for the wide band signals based on fourth order cumulants is presented. This algorithm is also applied to noisy data. Both the cumulants of the received data and multidimensional filtering processing are used to remove the additive noises. The principal interest of this preprocessing is the improvement of the signal to noise ratio at each analysis frequency. Then the signal subspace estimated from the average cumulant matrix resolves the fully correlated wide band acoustic sources. The simulation results are used to evaluate the performance and to compare the different coherent signal subspace algorithms. The obtained results show that the performances of the proposed algorithm are similar to those of the spectral matrix based methods when the noise is white and are better in the presence of Gaussian or an unknown noise. The Std variations and the localization results indicate that the whitening of the received data improves the localization in the presence of no completely Gaussian noise.

13. Acknowledgment

The authors would like to thank Dr Jean-Pierre SESSAREGO for providing real data and for his useful collaboration.

14. References

Reilly, J. & Wong, K. (1992). Estimation of the direction-of-arrival of signals in unknown correlated noise. Part II: Asymptotic behavior and performance of the MAP approach, *IEEE Trans. on signal processing*, Vol. 40, No. 8, (Aug. 1992) (2018-2028), ISSN 1053-587X .

Wu,Q. & Wong, K.(1994). UN-MUSIC and UN-CLE: An application of generated correlation analysis to the estimation of direction-of-arrival of signals in unknown correlated noise. *IEEE Trans. on signal processing*, Vol.42, No. 9, (1994) (2331-2343).

Stoica, P.; Vieberg, M. & Ottersten, B. (1994). Instrumental variable approach to array processing in spatially correlated noise fields. *IEEE Trans. on signal processing*, Vol.42, No. 1, (Jan. 1994) (121-133), ISSN 1053-587X .

Wax, M. (1991). Detection and localization of multiple sources in noise with known covariance. *IEEE Trans. on signal processing*, Vol. 40, No. 1, (Jan. 1992) (245-249), ISSN 1053-587X.

Ye, H. & De Groat, R. (1995). Maximum likelihood DOA estimation and asymptotic Cramer Rao Bounds for additive unknown colored noise. *IEEE on signal processing*, Vol. 43, No. 4, (Apr. 1995) (938-949), ISSN 1053-587X .

Zhang, Y. & Ye, Z. (2008). Efficient Method of DOA Estimation for Uncorrelated and Coherent Signals. *IEEE Antennas and Wireless Propagation Letters*, Vol. 7, (2008) (799-802), ISSN 1536-1225 .

Tayem, N.; Hyuck M. & Kwon. (2006). Azimuth and elevation angle estimation with no failure and no eigen decomposition. *EURASIP Signal Processing*, Vol. 86, No. 1, (2006) (8-16).

Cadzow, J.(1988). A high resolution direction-of-arrival algorithm for narrow-band coherent and incoherent sources. *IEEE Trans. on Acoustics, Speech and Signal Processing*, Vol. 36, No. 7, (1998) (965-979).

Werner, K. Jansson, M. (2007). DOA Estimation and Detection in Colored Noise Using Additional Noise-Only Data. *IEEE Trans. on signal processing*, Vol. 55, No. 11, (Nov. 2007) (5309-5322), ISSN 1053-587X .

Friedlander, B. & Weiss, A. (1995). Direction finding using noise covariance modeling. *IEEE Trans. on signal processing*, Vol. 43, No. 7, (Jul. 1995) (1557-1567), ISSN 1053-587X .

Abeidaa, H. & Delmas, P. (2007). Efficiency of subspace-based DOA estimators. *EURASIP Signal Processing*, Vol. 87, No. 9, (Sept. 2007) (2075-2084).

Kushki, A. & Plataniotis, K.N. (2009). Non parametric Techniques for Pedestrian Tracking in Wireless Local Area Networks. *Handbook on Array Processing and Sensor Networks* Haykin, S., Liu, K.J.R.,(Eds.), John Wiley & Sons, (783-803), ISBN 9780470371763

Hawkes, M. & Nehorai, A. (2001). Acoustic vector-sensor correlations in ambient noise. *IEEE J. Oceanic Eng.*, Vol. 26, No. 3, (Jul. 2001) (337-347).

Yuen, N. & Friedlander, B. (1997). DOA estimation in multipath : an approach using fourth order cumulants. *IEEE. Trans. on signal processing*, Vol. 45, No. 5, (May 1997) (1253-1263), ISSN 1053-587X .

Valaee, S. & Kabal, P. (1995). Wideband array processing using a two-sided correlation transformation. *IEEE Trans. on signal processing*, Vol. 43, No. 1, (Jan. 1995)(160-172), ISSN 1053-587X .

Hung, H. & Kaveh, M. (1988). Focusing matrices for coherent signal-subspace processing. *IEEE Trans. on Acoustics, Speech and Signal Processing*, Vol.36, No.8, (Aug. 1988) (1272-1281), ISSN 0096-3518 .

Bourennane, S.; Frikel M. & Bendjama, A. (1997). Fast wide band source separation based on higher order statistics. *Proceedings of IEEE Signal Processing Workshop on Higher Order Statistics*, pp.354-358, DOI 10.1109/HOST.1997.613546, Banff Canada, Jul. 1997.

Bendjama, A.; Bourennane, S. & Frikel, M. (1998). Direction Finding after Blind Identification of Source Steering Vectors. *Proceedings of SIP'98*, Las Vegas US, Oct. 1998.

Mendel, J.M. (1991). Tutorial on higher order statistics(spectra) in signal processing and system theory : Theoretical results and some applications. *IEEE Proceedings*, 1991, Vol.79, No.3, pp.278-305, March 1991.

Pillai S.& Kwon B. (1989). Forward/Backward spatial smoothing techniques for coherent signal identification. *IEEE Trans. on Acoustics, Speech and Signal Processing*, Vol.37, No.1, (jan. 1989)(8-15), ISSN 0096-3518 .

5

Narrowband Interference Suppression in Underwater Acoustic OFDM System

Weijie Shen, Haixin Sun*,
En Cheng, Wei Su and Yonghuai Zhang
The Key Laboratory of Underwater Acoustic Communication and Marine Information Technology, Xiamen University, Xiamen, Fujian, China

1. Introduction

In 1960, Chang [1] postulated the principle of transmitting messages simultaneously through a linear band-limited channel without ICI and ISI. Shortly after, Saltzberg [2] analyzed the performance of such a system and concluded, "The efficient parallel system needs to concentrate more on reducing cross talk between the adjacent channels rather than perfecting the individual channel itself because imperfection due to cross talk tends to dominate." This was an important observation and was proved in later years in the case of baseband digital signal processing.

The major contribution to the OFDM technique came to fruition when Weinstein and Ebert [3] demonstrated the use of the Discrete Fourier Transform (DFT) to perform the base-band modulation and demodulation. The use of the DFT immensely increased the efficiency of the modulation and demodulation processing. The use of the guard space and the raised-cosine filtering solve the problems of ISI to a great extent. Although the system envisioned as such did not attain the perfect orthogonality between sub-carriers in a time-dispersive channel, nonetheless it was still a major contribution to the evolution of OFDM systems.

In quest of solving the problem of orthogonality over the dispersive channel, Peled and Ruiz [4] introduced the notion of cyclic prefix (CP). They suggested filling the guard space with the cyclic extension of the OFDM symbol, which acts as if it is performing the cyclic convolution by the channel as long as the channel impulse response is shorter than the length of the CP, thus preserving the orthogonality of sub-carriers. Although addition of the CP causes a reduction of the data rate, this deficiency was more than compensated by the ease of receiver implementation.

Orthogonal Frequency Division Multiplexing (OFDM) scheme enables channel equalization in the frequency domain, thus eliminating the need for potentially complex time-domain equalization of a single-carrier system [5]. For this reason, OFDM has found application in a number of systems. Underwater acoustic(UWAC) OFDM system has been widely studied to overcome the complexity of underwater acoustic channels [6] and get a high speed data

---

* Corresponding Author

transmission. Nevertheless, due to the effect of"spectral leakage"in discrete Fourier transform(DFT)[7],which is used in OFDM receivers, spectrum of narrowband interference(NBI) will be spread in the whole frequency domain , thus, strongly impacts adjacent sub-channels [8-9].Various algorithms have been proposed for NBI suppression. Aiming at NBI suppression in OFDM systems, some approaches are employed. Receiver window method is well known as a simple and valuable NBI suppression technique [10]. Literatures [11] consider spreading the OFDM symbols over subcarriers using orthogonal codes. A linear minimum mean square estimator (LMMSE) for the NBI cancelation in frequency domain is introduced in [12]. Prediction error filter is introduced for mitigating the interference in the time domain, thereby preventing the spectralleakage of the interference power [13]. PEF is based on the fact that the OFDM signal can't be predicted since it has a flat spectrum, while the NBI can be predicted precisely. The effect of guard band in the PEF based NBI suppression is discussed, and two methods to solve this problem are discussed. It is shown that the proposed decimated prediction filter (DPF) and virtual subcarriers complement prediction error filter (VSC-PEF) outperform conventional PEF in OFDM systems with guard band [14]. Shi et. al [15] gave a comprehensive analysis of the impact of NBI on an OFDM based ultrawide- band (UWB) transceivers including ADC, timing and frequency synchronization. However, none has been done to examine how the NBI affects the cross-correlation based timing synchronization scheme [16], which has also been widely adopted. This paper examines this problem and provides theoretical analysis of timing synchronization of an OFDM system in the presence of multi-tone NBI. Numerical simulations are presented to verify the theoretical analysis. A Nyquist window is proposed for OFDM receiver without additional expense of system bandwidth, and choose the rectangular window and the raised cosine window window for analysis and simulation.The study was based on the experimental results in real underwater acoustic channel.[17]

This chapter is divided into five sections. Section 1.2 discusses the basic OFDM concept. The OFDM model is introduced in Section 1.3. Different kinds of window were described in Section 1.4, as well as Narrowband Interference Suppression methods in underwater acoustic communication systems. Simulation and experiment results were given in Section 1.5.

2. OFDM concept and its model

2.1 Evolution of OFDM

Frequency Division Multiplexing (FDM) has been used for a long time to carry more than one signal over a telephone line. FDM divides the channel bandwidth into sub-channels and transmits multiple relatively low rate signals by carrying each signal on a separate carrier frequency. To ensure that the signal of one sub-channel did not overlap with the signal from an adjacent one, some guard-band was left between the different sub-channels. Obviously, this guard-band led to inefficiencies.

In order to solve the bandwidth efficiency problem, orthogonal frequency division multiplexing was proposed, where the different carriers are orthogonal to each other. With OFDM, it is possible to have overlapping sub-channels in the frequency domain, thus increasing the transmission rate. The basis functions are represented in Figure 1. This carrier spacing provides optimal spectral efficiency.

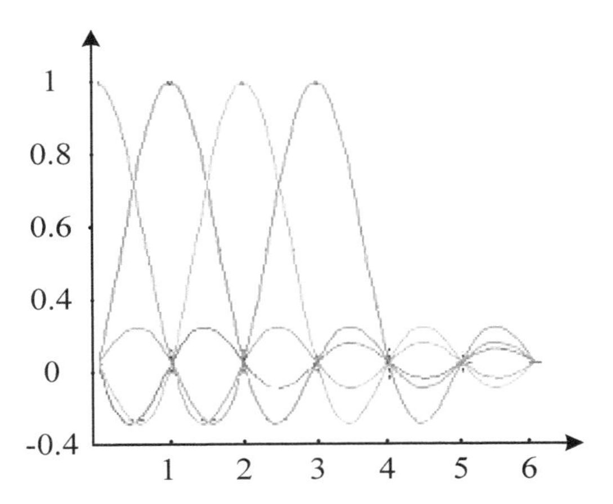

Fig. 1. Basis functions in OFDM system

Today, OFDM has grown to be the most popular communication system in high-speed communications.

2.2 Introduction to OFDM

The OFDM system studied in this paper has the block structure as shown in Figure 2. The system maps the input bits into complex-valued symbols $X(n)$ in the modulation block, which determines the constellation scheme of each sub-carrier. The number of bits assigned to each sub-carrier is based on the signal to noise ratio of each sub-carrier on the frequency range. The adaptive bit loading algorithm will be detailed below. The IFFT block modulates $X(n)$ onto N orthogonal sub-carriers. A cyclic prefix is then added to the multiplexed output of IFFT. The obtained signal is then converted to a time continuous analogy signal before it is transmitted through the channel. At the receiver side, an inverse operation is carried out and the information data is detected.

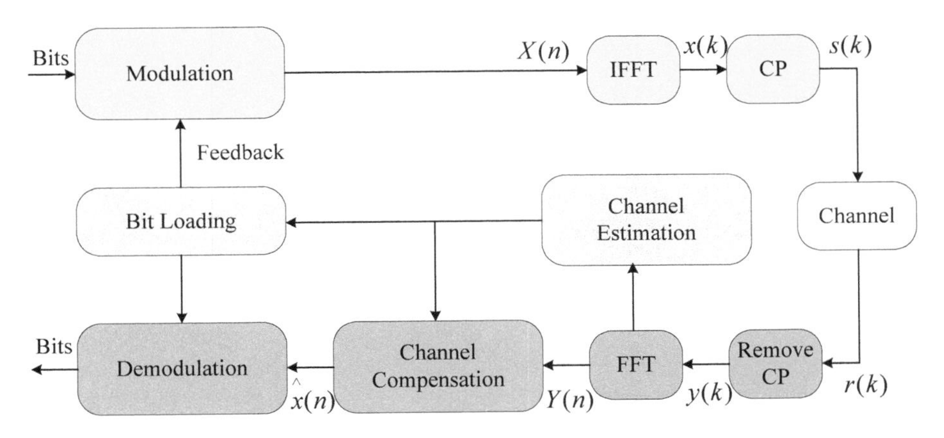

Fig. 2. OFDM system

2.3 FFT and IFFT

The key components of an OFDM system are the inverse FFT at the transmitter and FFT at the receiver. These operations performing linear mappings between N complex data symbols and N complex OFDM symbols, result in robustness against fading multi-path channel. The reason is to transform the high data rate stream into N low data rate streams, each experiencing a flat fading during the transmission. Suppose the data set to be transmitted is $X(1), X(2), ..., X(N)$, where N is the total number of sub-carriers. The discrete-time representation of the signal after IFFT is:

$$x(n) = \frac{1}{\sqrt{N}} \sum_{k=0}^{N-1} X(k) \cdot e^{-j2\pi k \frac{n}{N}}, \qquad n = 0 \ldots N-1 \tag{1}$$

At the receiver side, the data is recovered by performing FFT on the received signal,

$$Y(k) = \frac{1}{\sqrt{N}} \sum_{n=0}^{N-1} x(n) \cdot e^{-j2\pi k \frac{n}{N}}, \qquad k = 0 \ldots N-1 \tag{2}$$

An N-point FFT only requires $N \log(N)$ multiplications, which is much more computationally efficient than an equivalent system with equalizer in time domain.

2.4 Cyclic prefix

In an OFDM system, the channel has a finite impulse response. We note t_{max} the maximum delay of all reflected paths of the OFDM transmitted signal, see Figure 3.

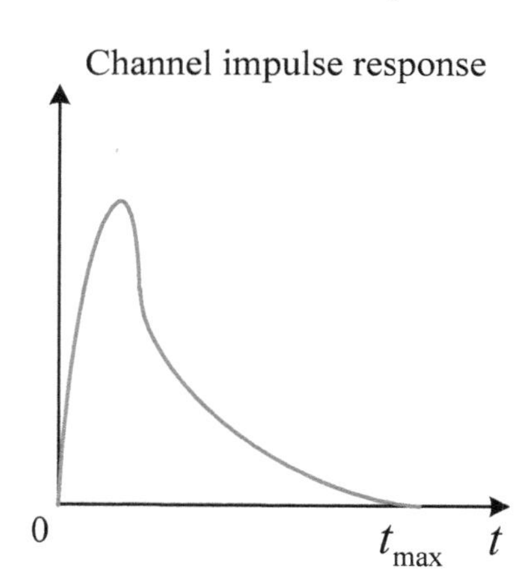

Fig. 3. Channel impulse response

Cyclic prefix is a crucial feature of OFDM to combat the effect of multi-path. Inter symbol interference (ISI) and inter channel interference (ICI) are avoided by introducing a guard interval at the front, which, specifically, is chosen to be a replica of the back of OFDM time domain waveform. Figure 4 illustrates the idea.

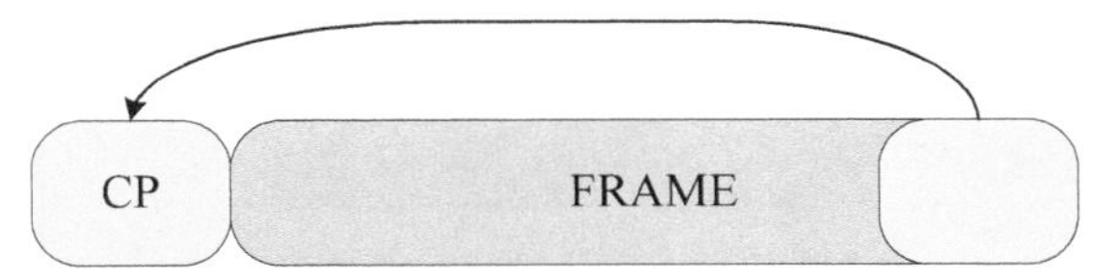

Fig. 4. Adding a cyclic prefix to a frame

From above expressions the sub-carrier waveforms are now given by

$$s(k)=\begin{cases} x(k+N) & -M \le k < 0 \\ x(n)=\frac{1}{\sqrt{N}}\sum_{k=0}^{N-1} X(k)\bullet e^{j2\pi k\frac{n}{N}} & 0 \le k < N-1 \end{cases} \tag{3}$$

The idea behind this is to convert the linear convolution (between signal and channel response) to a circular convolution. In this way, the FFT of circular convolved signals is equivalent to a multiplication in the frequency domain. However, in order to preserve the orthogonality property, $t_{\max}$ should not exceed the duration of the time guard interval. Once the above condition is satisfied , there is no ISI since the previous symbol will only have effect over samples within $[0,t_{\max}]$. And it is clear that orthogonality is maintained so that there is no ICI.

$$r(k)=s(k)\otimes h(k)+e(k) \tag{4}$$

$$\begin{aligned} Y(n) &= DFT(y(k)) = DFT(IDFT(X(n))\otimes h(k)+e(k)) \\ &= X(n)\bullet DFT(h(k))+DFT(e(k)) \\ &= X(n)\bullet H(n)+E(n) \qquad ,0 \le k \le N-1 \end{aligned} \tag{5}$$

where $\otimes$ denotes circular convolution and $E(n)=DFT(e(k))$. Another advantage with the cyclic prefix is that it serves as a guard between consecutive OFDM frames. This is similar to adding guard bits, which means that the problem with inter frame interference also will disappear.

2.5 Modulation and demodulation

Given the adaptive bit loading algorithm, the modulator has a number of bits and an energy value as input for each sub-carriers. The output for one sub-carrier is a constellation symbol with a desired energy, corresponding to the number of bits on the input. The modulator is taken to get either 2bits, 4bits, 6bits or 8bits available, which means that, respectively, only QPSK, 16QAM, 64QAM and 256QAM are available for modulation on each sub-carrier.

Demodulation is performed using Maximum Likelihood (ML) approach, given knowledge of the flat fading channel gain for each sub-carrier.

Furthermore, in order to reduce bit errors, Gray-coded constellations are also employed for each modulation order available. This Gray coding ensures that if a symbol error occurs, i.e. the decoder selects an adjacent symbol to what the transmitter intended to be decoded, there is only a single bit error resulting.

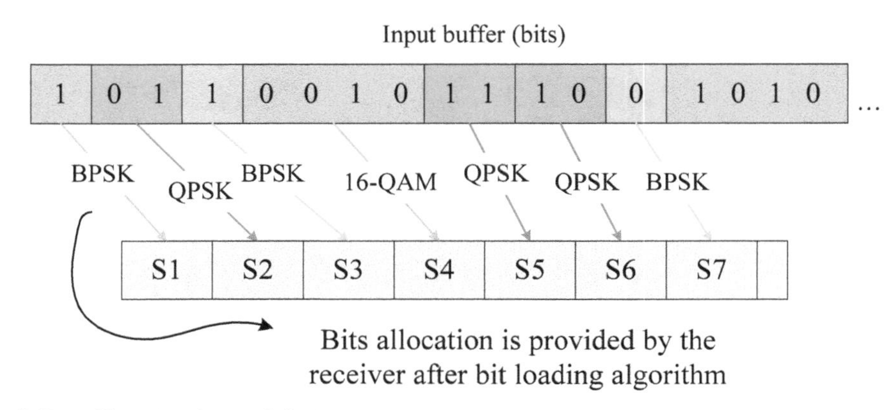

Fig. 5. Bits allocation in modulation

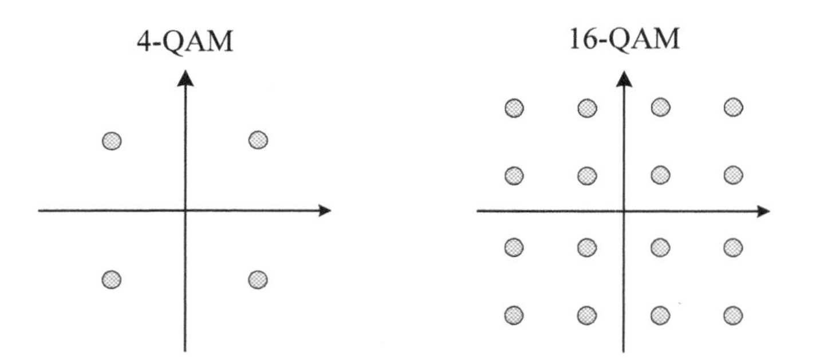

Fig. 6. 4-QAM(left) and 16-QAM(right)

Large performance improvement can be obtained in this adaptive modulation application where the modulator basis functions are designed as a function of measured channel characteristics. On good sub-channel (high SNR), modulation methods, such as 64 QAM, are used to increase the bit rate. And a lower modulation, such as QPSK, is performed on bad sub-channel to keep the error rate in a low level.

2.6 Interference

The interference signal selected is a non-orthogonal sinusoid relative to the OFDM waveform orthogonal basis set and has been added in the physical layer as part of the channel. A non-orthogonal interference signal has been selected because the zeros of the spectral response of the non-orthogonal interference signal do not fall on the zeros of the spectral response of the OFDM orthogonal basis set. As shown in figure 7, the difference in the spectral responses of the two types of interference signals measured at the zero crossings of the orthogonal basis set of the OFDM wave-form vary by 350 dB, which is limited by the computational precision of the algorithm. The result is the spectral side-lobes of the non-orthogonal interference signal corrupt each channel of the OFDM waveform, whereas the orthogonal interference signal corrupts only one channel.

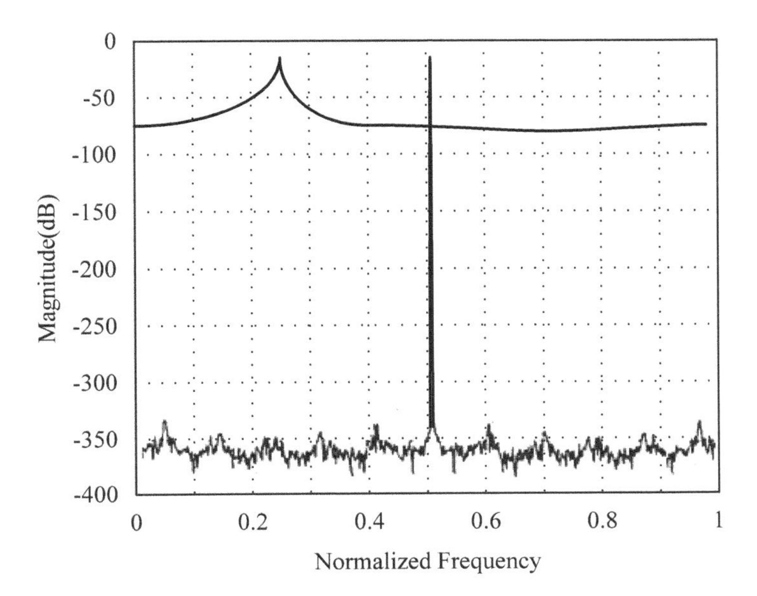

Fig. 7. Spectral Characteristics of the Interference Signal used to Demonstrate the Difference Between an Orthogonal and Non- Orthogonal Interference Signal

Figure 8 summarizes the simulation results for measuring the bit error rate vs. E_b / N_0 in the presence of the non- orthogonal interference signal. The reference BER vs. E_b / N_0 curve has no interference signal added in the physical layer. The interference signal amplitude was varied relative to the energy in one channel in the OFDM waveform to measure the degradation in performance. For a OFDM modulator implemented as a $2N$ point DFT, the energy in either the in-phase or the quadrature component of one channel is given by the expression:

$$E_S = \frac{1}{2N}\overline{a^2} \tag{6}$$

where $\overline{a^2}$ is the average one-dimensional constellation energy of the input source and N is the number of channels per OFDM waveform. It should be noted that practical OFDM waveforms usually have a minimum of 2048 channels to extend the OFDM waveform duration for protection against fading. For high values of E_b / N_0, simulations require a large number of Monte Carlo iterations in order to have results that are statistically significant. For this simulation 500 Monte Carlo iterations have been selected. For N equal to 32 and the average one-dimensional constellation energy of the input source equal to 0.43 for 64-symbol QAM, the average energy per channel is:

$$\overline{E_s} = \frac{0.43}{2(32)} = 6.7(10)^{-3} \tag{7}$$

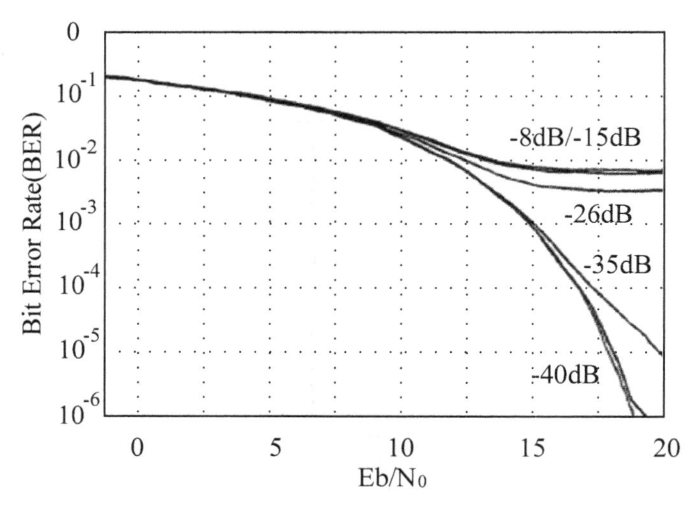

Fig. 8. Bit Error Rate vs. E_b / N_0 for Uncompensated OFDM with Varying Signal Energy from-8 dB to-40dB Relative to the Energy in one Channel.

As shown in figure 8, degradation in bit error rate is seen for interference signals that have signal energies larger than 40 dB below the energy in one channel of the OFDM wave-form. For interference signals with larger amplitude levels, the performance degradation quickly approaches a bit error rate of approximately $(10)^{-2}$ independent of E_b / N_0.

3. Narrowband interference suppression

3.1 Narrowband interference analysis

Narrow-band Interference are Single-frequency sinusoidal signals [18], which can be approximated as follows:

$$i(t) = a \cdot e^{(j2\pi f_c t+\theta)} \tag{8}$$

Where f_c is the interference frequency, and θ is a Random phase offset. Generally, the kth sample of the baseband received signal in time domain can be represented as

$$r(k) = \sum_{i=-\infty}^{+\infty} s(i) \cdot h(k-i) + n(k) + i(k) \tag{9}$$

where s is a diagonal matrix containing the transmitted signalling points, h is a channel attenuation vector, n is a vector of independent identically distributed complex zero-mean Gaussian noise with variance σ_n^2. The noise n is assumed to be uncorrelated with the channel h .And i is the interference. After FFT we got

$$R_{k,n} = H_{k,n} S_{k,n} + N_{k,n} + I_{k,n} \tag{10}$$

$R_{k,n}$, $H_{k,n}$, $S_{k,n}$, $N_{k,n}$ and $I_{k,n}$ represent the corresponding k^{th} frequency domain coefficient in the nth symbol. Where

$$I_{n,k} = \frac{\alpha \sin(\pi(\hat{f} N - K))}{N \sin(\pi(\hat{f} - K / N))} e^{j\pi(N+1)(\hat{f} - K/N)} \tag{11}$$

In the above equation, $\hat{f}$ is the normalize narrowband interference central frequency, $\hat{f} = f_c / f_s$, where f_s is the sampling frequency of OFDM symbols.

3.2 Leakage of narrowband interference and its suppression methods

3.2.1 The FFT filter bank

In the receiver, the analog OFDM signal passes the anti-alias filter before being sampled at a rate $f_a = 1 / T_a$. We assume the baseband signal is a complex low-pass signal without oversampling, therefore we obtain N complex samples each T_u and G samples each T_g. The kth sample $\underline{b}_k$ of a symbol is

$$\underline{b}_k = \sum_{i=1}^{N} \underline{c}_i e^{j2\pi f_i \cdot k \cdot T_a} \tag{12}$$

with $| f_i - f_{i-1} | = f_t$. The first step to recover carrier orthogonality is to eliminate for each symbol the G samples of the guard interval, as shown in figure 9. After that a discrete

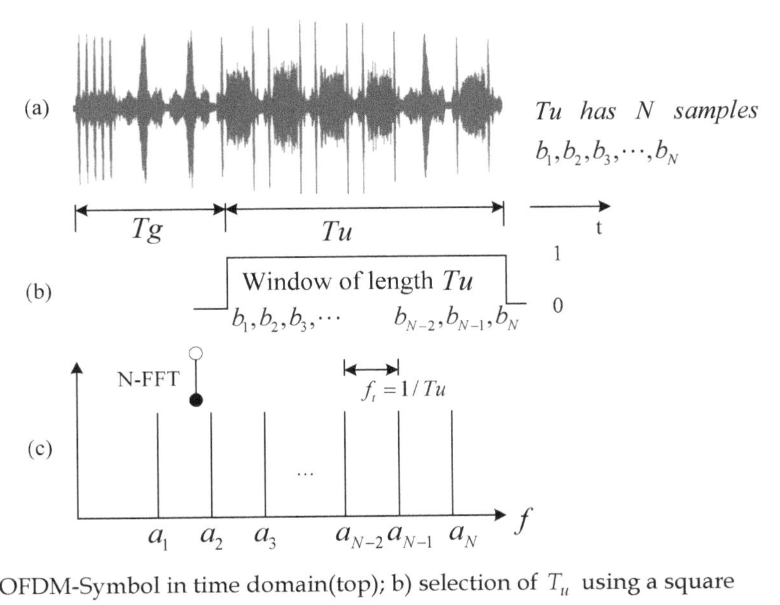

Fig. 9. a) OFDM-Symbol in time domain(top); b) selection of T_u using a square window(middle); c) OFDM-symbol in frequency domain(below)

Fourier transform DFT of the remaining N samples forces the periodicity of the time signal. The DFT performs the carrier filtering without inter-carrier-interference ICI, provided that $T_u = 1/f_t = NT_a$. The obtained complex coefficients a_n contain the original amplitude and phase information of the nth respective carrier. The orthogonaly filtered N carriers are shown in figure 9.

Using the proposal in [5] the filtering of each OFDM carrier is made with an FFT. The FFT has been considered in [19, 20] as a filter bank. The DFT $X(n)$ of a sampled signal $x(k)$ is defined

$$X(n) = \sum_{k=0}^{N-1} x(k) \bullet e^{-jkn\frac{2\pi}{N}} \tag{13}$$

Being F the normalized discrete frequency, whereby $F = N \cdot f / f_a$, the frequency response of the nth filter $H_n(F)$ is

$$|H_n(F)| = \left| \frac{\sin[\pi(F-n)]}{\sin[\pi(F-n)/N]} \right| \tag{14}$$

which is shown in figure 10 for $n = 16$. Due to the properties of the DFT this frequency response is periodical, with $|H_n(F-n)| = |H_n(F-n+N)|$. The integration of power components covered by the frequency response in figure 10 leads to the amplitude of FFT coefficient a_{16}.

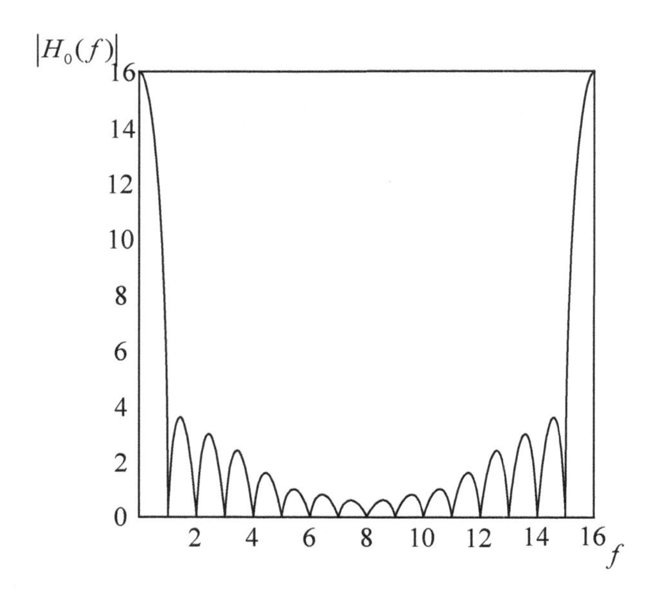

Fig. 10. Frequency response of a DFT filter ($n = 16$)

The reason why the DFT presents the frequency response in figure 10 is because the DFT performs the transform in blocks of only N samples, as shown in figure 9. This is equivalent to using a square window in time domain of length T_u corresponding to a $\sin(x)/x$ function in frequency domain. Due to the sampling in time domain, the frequency response is forced to periodicity leading to aliasing between the $si(x)$ functions. Therefore the (periodical) frequency response of the DFT is an aliased $si(x)$ function, also known as the Dirichlet function. Figure 11 shows the DFT filter bank consisting of N filters having the same shape of figure 10 but different center frequencies.

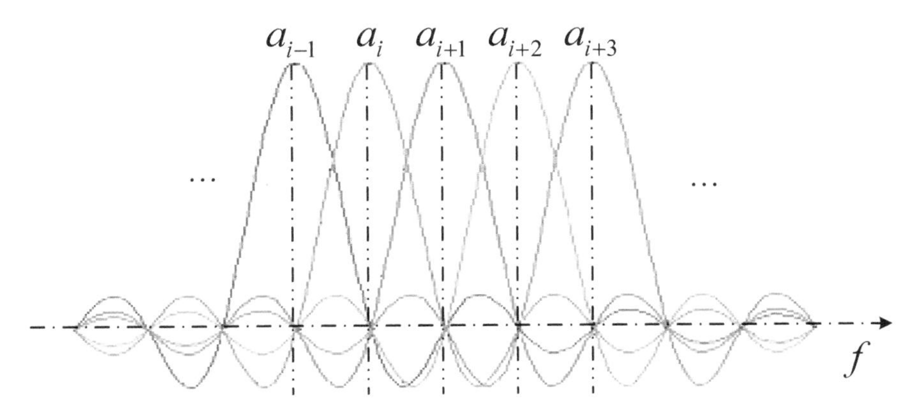

Fig. 11. Position of carriers in the DFT filter bank

The carriers are represented by ideal Dirac distributions placed on the filter maxima. The maximum of one filter coincides with the zero crossing of all others; this fact allows to separate the carriers without suffering any ICI. The interference occurs only for frequency components which does not correspond with the filter maxima.

In the non-ideal receiver the carriers are not exactly placed on the filter maxima due to frequency deviations producing the "leakage" of the DFT. This effect is explained in figure 12. The DFT of a single carrier without leakage ($F = 5$) consists of only one frequency component (FFT coefficient $a_5 = 16$), which includes the whole energy of the signal. In case of leakage ($F = 4.7$) the energy of the original carrier is distributed over all other frequency components (coefficients). The OFDM signal, consisting of many carriers which now "see" each other, is disturbed by IC1 due to leakage.

Leakage occurs also in case that the carriers are disturbed by phase noise leading to a widening of spectrum floor. White noise and disturbances like sine spurious also causes leakage. Nevertheless it is important to note that the leakage does not change the total signal power. The addition of the squared components (power) of figure 12 is always the same (256) independent of the carrier frequency [19, 20].

The DFT filter has the main disadvantage that its frequency response extends over the whole frequency range and presents high side lobes which decrease relatively slowly in frequency. Due to this fact the OFDM signal is very sensitive to frequency deviations and phase noise and discrete spurious. Therefore we need a filter bank with improved frequency response which keeps orthogonality that is easy to implement. The last requirements may be fulfilled using a nyquist windowing before computing the FFT.

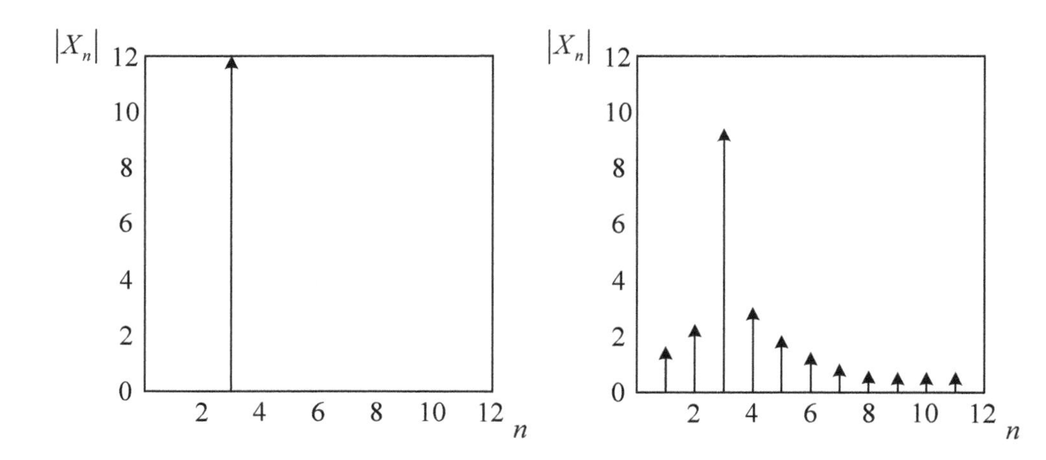

Fig. 12. DFT of a sine wave with leakage (left) and without leakage (right)

3.2.2 Using a Nyquist window in time domain

The use of a window e.g. Kaiser on the N samples before the FFT reduces the side lobe amplitude but also leads to an orthogonality loss between carriers. A windowing which reduces the side lobes and conserves the carrier orthogonality is called Nyquist window. It needs a repetition of every symbol in order to be implemented. For this purpose, the guard interval will be used. Since in many cases only a piece of the guard interval is necessary in order to avoid ISI, the other "usable" part of the guard interval may be used for the Nyquist windowing. Nevertheless $2N$ samples are necessary in order to maintain the orthogonality, as shown next.

The adaptive Nyquist windowing is able to improve the frequency response of the DFT filter. This windowing may be implemented with a guard interval length between 0 and T_u , $T_u \geq T_g \geq 0$. In case of a multi-path reception only the "usable guard interval" T_v (variable length) is used for the windowing. T_v is supposed to be free of ISI. The windowing is adaptive because T_v depends on modulation parameters and channel conditions. Figure 13 shows the way to implement an adaptive Nyquist windowing, as used in [21].

The usable part of the guard interval T_v, T_v consists of V samples. The V samples of T_v together with the OFDM-symbol T_u (N samples) are symmetrically windowed. The window $w(t)$ has a shape which fulfills the time domain analogy to the Nyquist criterion. Defining the "Nyquist time" T_u we expect to have no IC1 (in frequency domain!) if the window shape is symmetrical to T_u. In the same manner as for a pulse shaping filter, we may choose the Nyquist window using a raised cosine function, shown in figure 14.

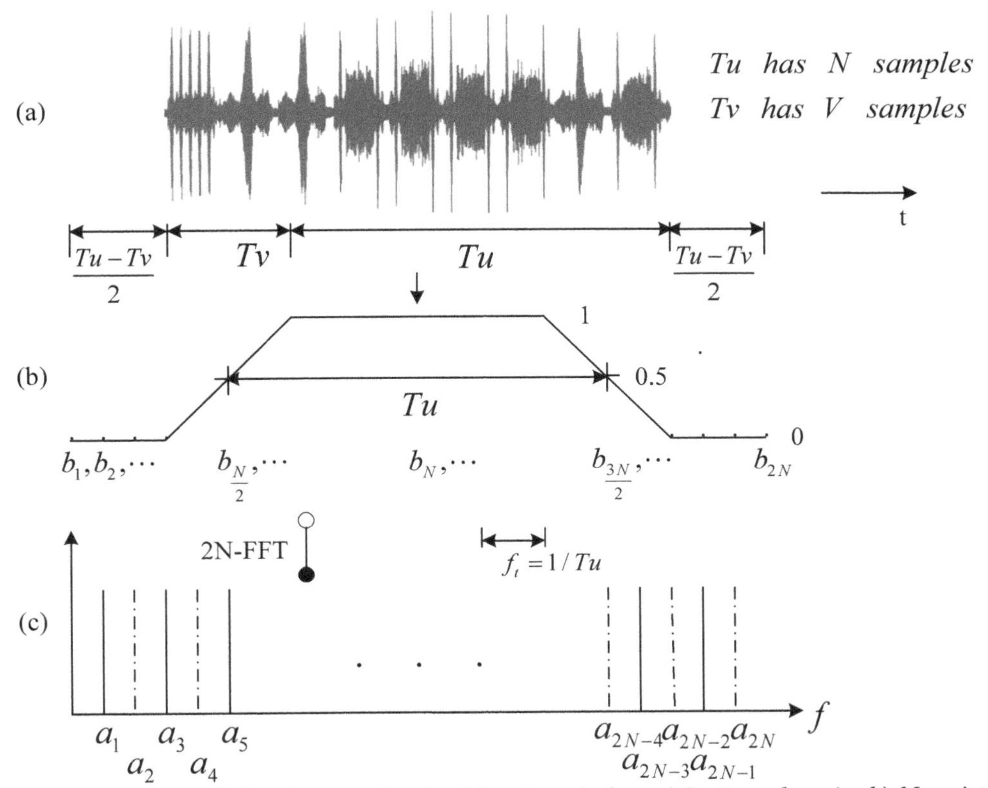

Fig. 13. OFDM-symbol using an adaptive Nyquist window a) In time domain, b) Nyquist window, c) Spectrum

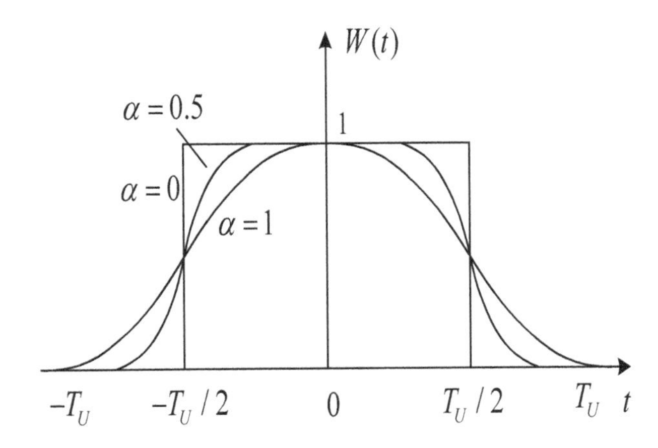

Fig. 14. Raised cosine window function w(t) depending of roll-off factor α

$$w(t)=\begin{cases}1 & 0\le|t|\le\dfrac{(1-\alpha)T_u}{2}\\ \dfrac{1}{2}\left[1-\sin\dfrac{\pi(|t|-T_u/2)}{\alpha T_u}\right] & \dfrac{(1-\alpha)T_u}{2}\le|t|\le\dfrac{(1+\alpha)T_u}{2}\\ 0 & |t|\ge\dfrac{(1+\alpha)T_u}{2}\end{cases} \tag{15}$$

We define a roll-off factor $\alpha = T_v / T_u$, which varies depending on the usable part of the guard interval. Practical values of α between 0 and 0.3 may be considered.

After Nyquist windowing the filter bank has lost its orthogonality, since the zero crossings are separated by $1/(T_u + T_v) \neq f_{tr}$. Neither a FFT of the total windowed samples $V+N$ is possible, because their number is not a power of two. Thus a symmetrical "zero padding" is performed in order *to* complete a total of $2N$ samples $b_{k,w}$, where

$$b_{k,w}=\begin{cases}0 & 1\le k<(N-V)/2\\ b_k w(t) & (N-V)/2\le k\le 3(N+V)/2\\ 0 & 3(N+V)/2<k<2N\end{cases} \tag{16}$$

In this way a FFT may still be used instead of the more complex non-power-of-two-DFT and the filter bank orthogonality is restored. The FFT computes the $2N$ coefficients α_k on the windowed samples $b_{k,w}$. Figure 15 shows the 2N-DFT filter frequency response depending on the roll-off factor.

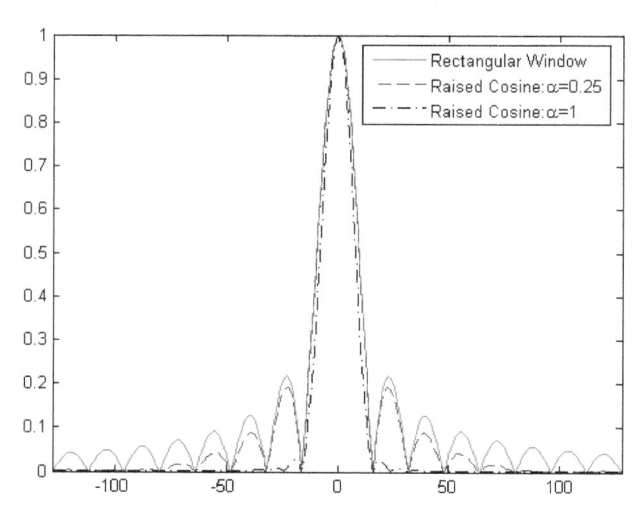

Fig. 15. Frequency response of 2N-DFT Filter with raised cosine windowing depending of roll-off factor α

The improvement of the frequency response occurs for $\alpha > 0$. The bigger α turns, the lower are the frequency response side-lobes. The area covered by the filter response is smaller and the C/N improves. The whole filter bank is less sensitive to frequency deviations, disturbances, etc. The reason for the improvement can also be explained through a decrease of the DFT-leakage. Since the leakage is responsible in several cases for an OFDM signal degradation, an overall improvement in demodulation is expected.

4. Simulation and experiment results

4.1 Experiment design

Figure 16 illustrates the block diagram of the underwater acoustic OFDM system[17]. The input data is first mapped into a QPSK constellation. Then the data sequence is converted in parallel and entered the FFT to perform spreading. Pilot signals are inserted before the IFFT operation which is used to implement OFDM modulation. A cyclic prefix is also appended to the data sequence as guard interval. The complex base-band digital to analog signal is then up-converted to the transmission frequency and sent out to the underwater acoustic channels by the transducer.

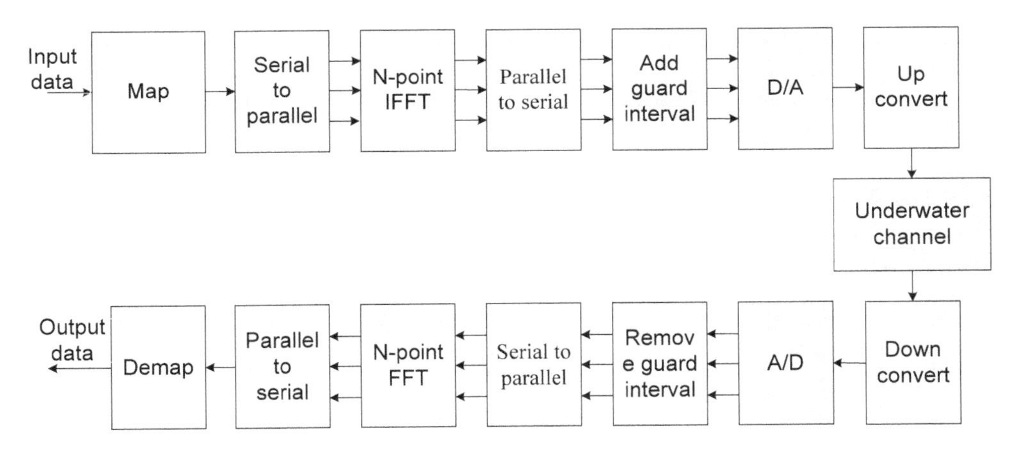

Fig. 16. Underwater acoustic OFDM system

In order to know the frame boundary, LFM signals are appended before the data sequence before transmitted into the channel. At receiver side, time synchronization is done via correlating the received samples with the known LFM sequence. After that, the received data are divided into OFDM demodulation. System specification for the experiment is shown in Table 1.

Table 2. shows the BER results of simulation with SNR=12dB for AWGN channel. The number of multipath interference is five. Each subcarrier in OFDM signal is QPSK mapped. The subcarrier number is 1024.

Mapping mode	QPSK
Bandwidth	6kHz
Carrier frequency	30kHz
Transmission frequency band	27~33kHz
Symbol duration	117.3ms

Table 1. System specification

	Rectangular Window	**Raised Cosine**
1	0.1092	0.0778
2	0.1001	0.0706
3	0.1056	0.0691
4	0.1047	0.0721
5	0.1062	0.0730
AVERAGE	0.1051	0.0725

Table 2. Simulation results

4.2 Experimental pool results

The experiment was carried out at the experimental pool in Xiamen University. The distance between the transmitter and the receiver is 10m.Both of them kept still during the whole experiment. The BER results of experiment were given in the Table 3, as well as the average BER. Figure 17 shows the reconstructed images using different window function in pool experiments.

	Rectangular Window	**Raised Cosine**
1	0.0799	0.0500
2	0.0791	0.0504
3	0.0773	0.0505
4	0.0781	0.0503
5	0.0785	0.0509
AVERAGE	0.0785	0.0504

Table 3. Experiment results

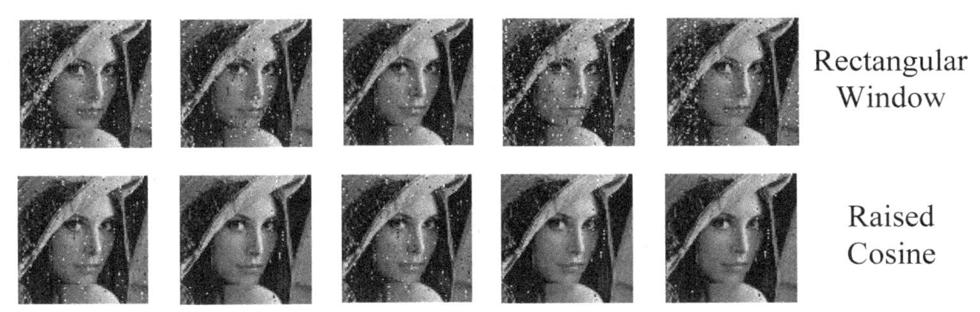

Fig. 17. Reconstructed images with different window function in pool experiments

4.3 Experimental results

The experiment was carried out in shallow water near Xiamen. Figure 18 depicts the location of transmitter and receiver transducers. The distance between transmitter and receiver was 810m. The depth of water was nearly 4m, and the transmitter and receiver were in the middle. According to the experiences before,the threshold T is set 2.6. Table 4 is the result of the experiment. The BER results were given in the table, as well as the average BER. Figure 19 shows the reconstructed images using different window function in shallow water experiments.

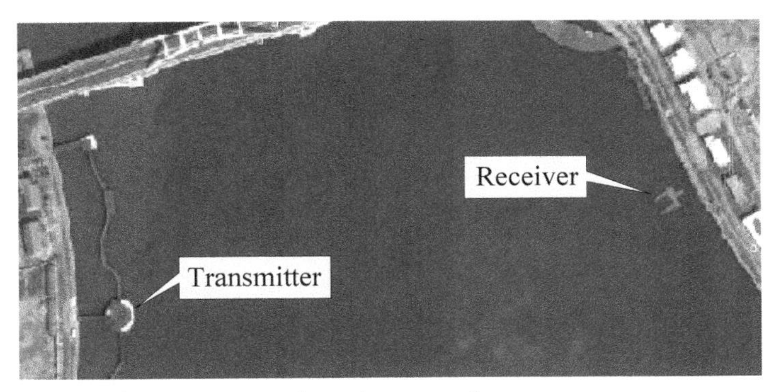

Fig. 18. The location of transmitter and receiver transducers

	Rectangular Window	Raised Cosine
1	0.1543	0.1057
2	0.1357	0.1103
3	0.1552	0.1034
4	0.1406	0.0998
5	0.1468	0.1028
AVERAGE	0.1465	0.1044

Table 4. Experiment results

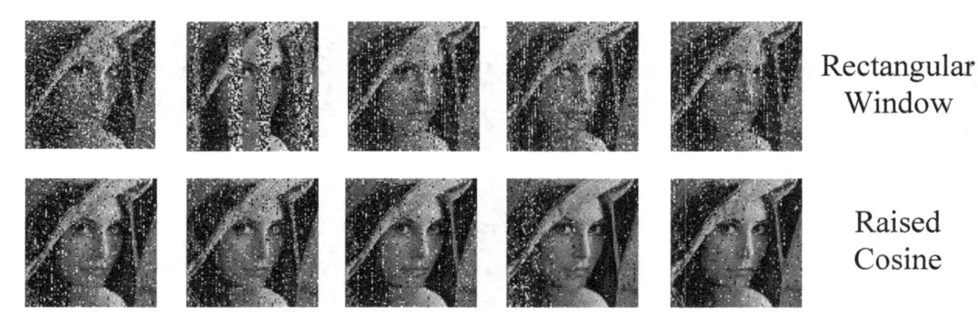

Fig. 19. Reconstructed images with different window function in shallow water experiments

5. Conclusion

We propose a Raised Cosine Window for OFDM receiver and apply it to Underwater Acoustic OFDM System. Based on the results of simulation and experiment, the Raised Cosine Window applied to OFDM receiver can suppress NBI and reduce the BER, which has simple structure, and only needs a little modification in the receiver.

The proposed algorithm holds three main advantages with respect to narrowband interference suppression in underwater acoustic OFDM system. First and foremost, the simulation and the experiment results in a shallow underwater channel showed that the raised cosine window based method could achieve excellent performance. Besides, the proposed method performed comparable bandwidth efficiency to other methods. Last but not least, it need not change the frame structure of existing protocol, which is vital for the detection methods that are exploited at the receiver. The study demonstrated that the raised cosine window based narrowband interference suppression would be applicable for OFDM system over underwater acoustic channels.

6. References

[1] Chang, R. W., "Synthesis of Band-Limited Orthogonal Signals for Multichannel Data Transmission," Bell Systems Technical Journal, Vol. 45, December 1960, pp. 1775–1796.

[2] Saltzberg, B. R., "Performance of an Efficient Parallel Data Transmission System," IEEE Trans. on Communications, Vol. COM-15, No. 6, December 1967, pp. 805–811.

[3] Weinstein, S. B., and P. M. Ebert, "Data Transmission of Frequency Division Multiplexing Using the Discrete Frequency Transform," IEEE Trans. on Communications, Vol. COM-19, No. 5, October 1971, pp. 623–634.

[4] Peled, A., and A. Ruiz, "Frequency Domain Data Transmission Using Reduced Computational Complexity Algorithms," Proc. IEEE Int. Conf. on Acoustics, Speech, and Signal Processing (ICASSP '80), Denver, CO, 1980, pp. 964–967.

[5] S. B. Weinstein and P. M. Ebert, "Data transmission by frequency-division multiplexing using the discrete fourier transform," IEEE Trans. Commun. , vol. 19, pp. 628–634, Oct. 1971.

[6] D. Brady and J. C. Preisig. Underwater Acoustic Communications. In H. V. Poor and G. W. Wornell, editors, Wireless Communications: Signal Processing Perspectives, chapter 8, pages 330–379. Prentice-Hall, 1998.

[7] Arthur J Redfem, "Receiver Window Design for Multicarrier Communication Systems", IEEE Journal on Selected Areas in Communications, vol.20, no.5, Jun 2002.

[8] Hermann Rohling, "Narrow Band Interferences Reduction in OFDM based Power Line Communication Systems", International Symposium on Power Line Communications (ISPLC) 2001.

[9] Zhang, D., Pingyi Fan, Zhigang Cao, "Receiver window design for narrowband interference suppression in IEEE 802.11a system",Proceedings. The 2004 Joint Conference of the 10th Asia-Pacific Conference on Volume 2, 29 Aug.-1 Sept. 2004 Page(s):839-842 vol.2

[10] A. J. Redfern, "Receiver window design for multicarrier communication systems," IEEE J. Sel. Areas Commun., vol. 20, pp. 1029–1036, 2002.

[11] Z. Wu and C. R. Nassar, "Narrowband interference rejection in OFDM via carrier interferometry spreading codes," IEEE Trans. Wireless Commun.,vol. 4, pp. 1491–1505, 2005.

[12] D. Gerakoulis and P. Salmi, "An interference suppressing OFDM system for wireless communication," in Proc. IEEE ICC, 2002, pp. 480–484.

[13] R. Nilsson, F. Sj¨oberg, and J. P. LeBlanc, "A rank-reduced LMMSE canceller for narrowband interference suppression in OFDM-based systems," IEEE Trans. Commun., vol. 51, pp. 2126–2140, 2003.

[14] A. Batra and J. R. Zeidler, "Narrowband interference mitigation in OFDM systems," in Proc. IEEE MILCOM, 2008, pp. 1–7.

[15] K. Shi, Y. Zhou, B. Kelleci, T. Fischer, E. Serpedin, and A. Karsilayan, "Impacts of narrowband interference on OFDM-UWB receivers: Analysis and mitigation," IEEE Transactions on Signal Processing, vol. 55, no. 3, pp. 1118– 1128, Mar. 2007.

[16] K. Yip, Y. Wu, and T. Ng, "Timing-synchronization analysis for IEEE 802.11a wireless LANs in frequency-nonselective Rician fading environments," IEEE Transactions on Wireless Communications, vol. 3, no. 2, pp. 387-394, Mar. 2004.

[17] Weijie Shen; Haixin Sun; En Cheng; Wei Su; Yonghuai Zhang; "Narrowband interference suppression in underwater acoustic OFDM system". Informatics in Control, Automation and Robotics (CAR 2010) 2nd International Asia Conference on , Publication Year: 2010 , Page(s): 496-499.

[18] Hermann Rohling, "Narrow Band Interferences Reduction in OFDM based Power Line Communication Systems", International Symposium on Power Line Communications (ISPLC) 2001.

[19] C. Muschallik, "EinfluR der Oszillatoren im Frontend auf ein. OFDM-Signal", Fernseh und Kinotechnik, April 1995.

[20] C. Muschallik, "Influence of RF Oscillators on an OFDM Signal", IEEE Transactions on Consumer Electronics, Vol. 41, No. 3, August 1995, pp. 592-603.

[21] C. Muschallik, V. Armbruster., "Verfahren und Schaltungsanordnung zur Verbesserung des Empfangsverhaltens bei der Ubertragung von digitalen Signalen", German Pat. appl., AZ P195 20 353.4, June 1995.

6

Iterative Equalization and Decoding Scheme for Underwater Acoustic Coherent Communications

Liang Zhao and Jianhua Ge
State Key Laboratory of Integrated Service Networks, Xidian University, Xi'an, P.R.China

1. Introduction

Digital communications through underwater acoustic (UWA) channels differ from those in other media, such as radio channels, due to the high temporal and spatial variability of the acoustic channel which make the available bandwidth of the channel limited and dependent on both range and frequency. In order to overcome disadvantage factors and maximize performance to conduct real-time information understanding, underwater acoustic communications require the higher degree of information extraction and development from all kinds of onboard acoustic sensors and processing systems. A higher performance communication technology is needed in order to focus high-performance data processing on the problems and tasks faced by human operators and decision-makers. In order to establiseh reliable data commnication on the severely band-limited underwater acoustic channels, bandwidth-efficient modulation techniques (i.e. coherent communications) should be employed to overcome the inter-symbol interference (ISI) caused by channel multi-path propagation. The effective approach to eliminate the ISI caused by multipath propagation is that adaptive decision feedback equalizer (ADFE) integrates with spatial diversity. That is multi-channel adaptive decision feedback equalizer, which is applied in (Kilfoyle & Baggeroer, 2000; Stojanovic, 1996, 2005; Zhao et al., 2008) represents a more general approach to spatial and temporal signal processing.

However, Single technique, such as equalizer, is difficult to obtain satisfied data transmission because of the complexity of UWA channel, especially in shallow water channel. In recent years, more and more attention has been paid to Turbo codes, including parallel concatenated convolutional code (PCCC) (Berrou et al., 1993) and serially concatenated convolutional code (SCCC) (Benedetto & Montorsi, 1996), because of its near-capacity gains. The range of applications of Turbo codes has expanded to many areas of communications. Trellis-Coded Modulation (TCM) (Ungerboeck, 1982) is a kind of design option combining coding with modulation. It can provide over 3 dB coding gain without bandwidth expansion. Especially, it is interesting to combine PCCC or SCCC with TCM in order to improve the transmission spectral efficiency (i.e. parallel concatenated trellis codes modulation (PCTCM) (Benedetto et al., 1996; Chung & Lou, 2000; Legoff et al., 1994; Yang & Ge, 2005) and serial concatenated trellis codes modulation (SCTCM)) (Benedetto et,al., 1997; Divsalar & Pollara, 1997; Ho, 1997; Shohon et al., 2003) in order to improve the transmission spectral efficiency.

Therefore, iterative equalization and decoding (IED) based on equalizer and deocoding has been developed to obtain higher performance data transmission. Turbo equalizer (Berthet, 2000; Koetter et al., 2004) treats the channel encoder and channel itself as a serial concatenated system that can be decoded in an iterative scheme. A drawback of this iterative receiver is that the complexity of the turbo equalizer is orders of magnitude greater than the decision feedback equalizer (DFE). The turbo equalizer complexity grows exponentially with channel memory length. It isn't suit for the underwater acoustic channel with long delay spreads. In the structure of iterative euqalization and decoding, equalizer can use ADFE (Choi, 2008; Noorbakhsh et al., 2003) or equalizer based on channel estimation (Flanagan & Fagan, 2007; Otnes & Tuchler, 2004; Tuchler et al., 2002). For underwater acoustic channel with severely mutipath propagation and large time delay, adptive channel tracking can get better performance than channel estimation using trainning sequence or pilot symbols.

Comparing PCTCM, SCTCM has the following advantages: (1) It can further reduce the error floor of PCTCM to obtain lower BER (Soleymani & Gao, 2002). (2) It has more flexible coding structure than PCTCM. In this chapter, SCTCM technique is adopted to increase bandwidth efficiency. Furthurmore, a rate R=1 recursive convolutional is adopted as inner encoder of SCTCM encode to get higher performance SCTCM scheme. Therefore, iterative equalization and decoding, based on multi-channel adaptive decision feedback equalizer with variant step tracking factor and decoder of SCTCM, is formed to aid weight update of equalizer utilizing decoding gain provided by decoder of SCTCM such that the performance of equalizer is enhanced. And then, the performance of communication system is improved greatly through iteration calculation between equalizer and decoder.

The structure of this chapter is as follows. Firstly an overview of the channel and system model is provided in Section 2. More specifically, the channel model based on sound speed profile (SSP) measured in the lake and Bellhop method and the system description are discussed in Sections 2.1 and 2.2, respectively. And then, the introduction of the proposed iterative equalization and decoding is presented in Section 3. More specifically, the structures of iterative equalization and decoding are discussed in Section 3.1. The proposed iterative equalization and decoding process is detailed in Section 3.2, commencing with a discussion of the multichannel adaptive equalizer structure in Section 3.2.1, followed by a description of the SCTCM decoding algorithm in Section 3.2.2, the method of the soft symbol estimation in Section 3.2.3. Our simulation results are provided in Section 4, while Section 5 concludes our findings.

2. Channel and system models

We begin with the channel model under consideration and then discuss the description for communication systems.

2.1 Underwater acoustic channel model

We adopt real measured data, sound speed profile (SSP), and a finite-element ray (FER) tracing method (Bellhop) (Porter & Liu, 1994) to model the underwater acoustic multipath propagation. Additionally we model the multipath components as fading due to acoustic propagation loss.

A given multipath arrival l is characterized by its magnitude gain γ_l and delay τ_l. These quantities are dependent on the ray length l_l, which in turn is a function of the given propagation range R. The path magnitude gain is given by

$$\gamma_l = \frac{\Gamma_l}{\sqrt{\beta(l_l)}} \tag{1}$$

where, Γ_l is the amount of loss due to reflection at the bottom and surface. The acoustic propagation loss, represented by $\beta(l_l)$

$$\beta(l_l) = l_l^{\alpha}\left[a(f_c)\right]^{l_l} \tag{2}$$

where, α is constant, f_c is the carrier frequency and absorption coefficient (in db/km) given by Thorp's formula

$$\begin{aligned} &10\log a(f_c) \\ &= \frac{0.11 f_c^2}{1+f_c^2} + \frac{44 f_c^2}{4100+f_c^2} + 2.75\times10^{-4} f_c^2 + 0.003 \end{aligned} \tag{3}$$

The path delay is given by

$$t_l = \sum_{i=1} \frac{l_{l,i}}{v_{l,i}} \tag{4}$$

where, $v_{l,i}$ is the sound speed of the ith water layer according to SSP. Thus, the overall channel impulse response is given by

$$h(t) = \sum_{l=1}^{L} A_l \delta(t-\tau_l) \tag{5}$$

where, L is the multipath number, A_l and τ_l are the amplitude and relative delay of the lth multipath arrival respectively. $\tau_l = t_l - t_{\min}$, $t_{\min}$ is the minimus delay among the all path delays.

In the simulation section (Section 4), the SSP, measured on the lake (shown in Fig. 1), is adopted to model the multipath propagation. The SSP denotes the sound speed is changed with water depth.

Additionally, Doppler frequency f_d is considered in the channel model. It is given by

$$f_d = f_c \frac{v_r}{c} \cos\theta_l \tag{6}$$

Where, c donotes underwater sound speed, v_r is the relative speed between tranmiter and receiver, θ_l is the arrival angle for the lth arrival ray.

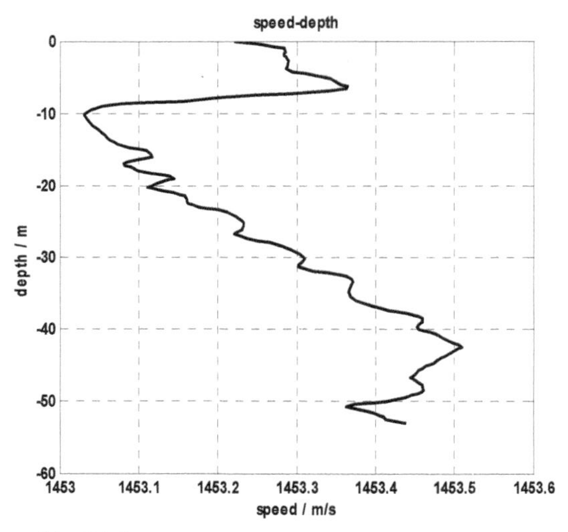

Fig. 1. Sound speed profile (SSP)

2.2 System description

The structure of transmitter is shown in Fig.2.

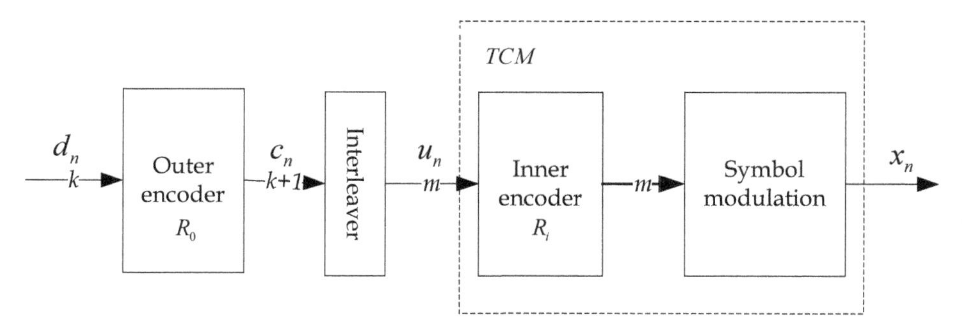

Fig. 2. The structure of transmitter

In comparison with high performance PCTCM scheme (Robertson & Woerz, 1998), the method in (Divsalar et al., 2000), with lower complexity, is adopted to design SCTCM, which can achieves $km/(k+1)\ bit/s/Hz$, using a rate $R_0 = k/k+1$ convolutional encoder with maximum free hamming distance as the outer code. An interleaver permutes the output of the outer code. The interleaved data enters a rate $R_i = m/m = 1$ recursive convolutional inner encoder. The m output bits are then mapped to one symbol belonging to a 2^m level modulation. In our system, the data symbol is QPSK modulated ($m = 2$), i.e. $x_n = \pm\frac{\sqrt{2}}{2} \pm \frac{\sqrt{2}}{2}$ with probability $\frac{1}{4}$. Before data symbol, the pilot symbol is transmitted to probe the channel impulse response (CIR). The LFM signal is used in our system. The frame structure is shown in Fig.3.

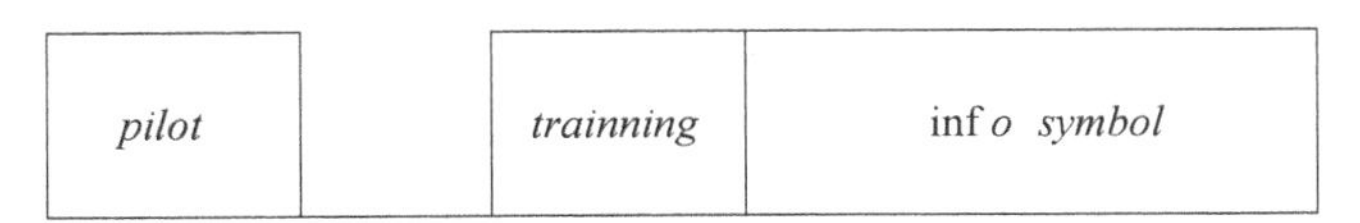

Fig. 3. Data frame structure

In receiver, Spatial diversity is achieved via multiple receiver arrays. The received signal at the kth array is given by

$$r_k(t) = \sum_{l=1}^{L} A_l x(t-\tau_l) e^{j2\pi(1+\Delta_l)f_c t} + n(t) \tag{7}$$

where, $\Delta_l = \frac{v_r}{c}\cos\theta_l$ is the Doppler frequency factor for lth multipah propagation, $n(t)$ is assumed to be a white Gaussian process with zero mean and variance σ_n^2.

After demodulation by multiplying the local carrier frequency and Doppler frequency compensation, the received baseband signal at the kth array is given by

$$y_k(t) = \sum_{l=1}^{L} A_l x(t-\tau_l) e^{j\theta_k} + n(t) \tag{8}$$

where, θ_k is the remain phase distoration.

And then, the iterative equalization and decoding (IED) is performed on the received multichannel baseband signals. The concrete IED algorithm will be presented in the next section.

3. Iterative equalization and decoding (IED)

In this section, we first present the structures of iterative equalization and decoding and analyze the merits and drawbacks of different structures in Section 3.1. And then, the proposed iterative receiver is detailed in Section 3.2.

3.1 The structures of iterative equalization and decoding

Since the underwater acoustic ISI channel can be treated as a convolutional encoder with rate 1, it is possible to treat the channel encoder and channel itself as a serial concatenated system that can be decoded in an iterative scheme such as the turbo equalizer structure that is illustrated in Fig.5. The motivation for the study of this receiver algorithm is to improve equalizer performance beyond that attainable by the optimum parameters decision feedback equalizer which also employs the all-training sequence.

The turbo equalizer consists of two soft input soft output (SISO) modules for the channel equalizer and the decoder that are arranged in a serial fashion. A drawback of this receiver algorithm is that the complexity of the turbo equalizer is orders of magnitude greater than the DFE. The turbo equalizer complexity grows exponentially with channel memory length, modulation level, and spatial diversity combining. It should be noted that traditionally the turbo equalizer has been used for known channels with reasonable ISI. It still needs to be demonstrated that this type of receiver can be used with modification to track the time-varying underwater channel and provide performance that exceeds the performance of the DFE using known training sequence throughout the entire data packet.

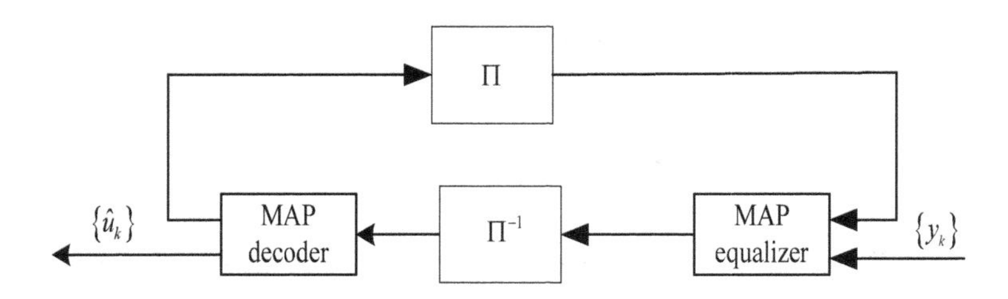

Fig. 4. Turbo equalizer

The second structure is hard iterative (shown in Fig.5.)

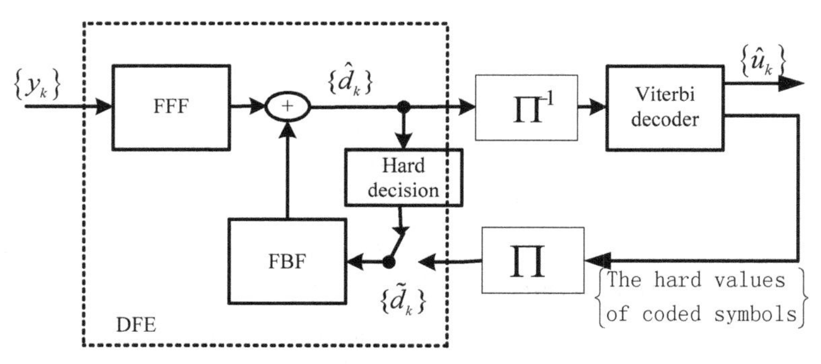

Fig. 5. Hard iteration

In this structure, the decoded symbols or hard decisions from the decoder after a first pass through the receiver system are then re-encoded to be used as the new training sequence to be used over the entire received data packet. Therefore, in the second pass or iteration, decision directed equalization after the short initial training sequence is not employed. Subsequent passes or iterations through the data in this fashion can be made. However diminishing improvements are obtained due to the hard decision nature of this algorithm. Feedback error propagation can still occur in this algorithm due to uncorrected errors at the output of the soft input hard output decoder. The desire for performance improvement by using SISO decoders as well as added information that the equalizer can provide to correct decoder errors provides the motivation for the soft iterative approach.

An improved receiver algorithm as compared to the hard iterative approach would be to employ all the information regarding the received symbols to generate the new training sequence by combining soft values of the coded symbols out of the decoder and the soft information about the detected symbols provided by the decision directed mode of the equalizer. This is intent of the soft iterative manner.

In the soft iteration, the decision feedback equalizer is modified so that it can use soft a priori information from the decoder from previous iterations. In order to obtain soft a priori information as opposed to hard decisions, the decoder structure must now take the form of a SISO device. One such device is the Maximum A posteriori (MAP) decoder. The method in which the information streams from the decoder and equalizer are combined is crucial because log likelihood ratio (LLR) values are produced from the decoder feedback path. In this chapter, the soft iteration structure is adopted.

3.2 Soft iterative equalization and decoding (IED)

The structure of the proposed IED with phase compensation is shown in Fig.6.

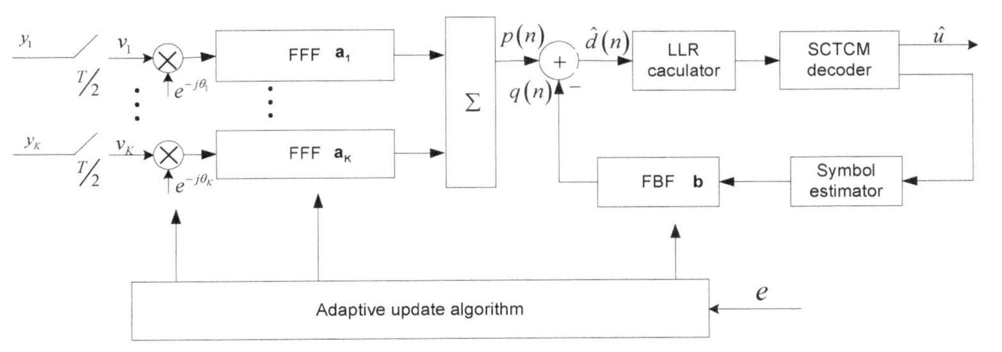

Fig. 6. The proposed IED

As shown in Fig.6, the received signal at each array elements is T/2 fractional sampled. The digital phase lock loop (DPLL) is adopted to correct phase distortion. Then, the feedforward and feedback filters are applied to obtain the estimation of transmitted symbol. In the IED scheme, the multichannel adaptive decision feedback qualizaer with phase compensation and decoder of SCTCM exchange soft information in an iterative manner. Specifically, at the output of the equalizer, the likelihood ratio (LLR) calculator computes soft information of coded bits based on the symbol estimation $\hat{d}(n)$. This soft information is delivered to the maximum a posteriori (MAP) decoder of SCTCM. In addition to providing the decoded output, the decoder also computes soft information on the coded bits, which is converted to soft estimates of the symbols. These soft symbol estimates are used to aid the operation of the equalizer and its adaptive weight update algorithm.

3.2.1 Mutichannel adaptive equalization

The received signal will carry out equalization proccessing (including carrier phase compensation) after demodulation. The main task of multichannel adaptive decision feedback qualizaer is eliminate inter-symbol interference (ISI) caused by multipath propagation.

According to minimum mean square error (MMSE) scheme, an error signal is used to update receiver parameters. The error signal $e(n)$ of adaptive update algorithm as follows

$$e(n) = d(n) - \hat{d}(n) \tag{9}$$

$$\hat{d}(n) = p(n) - b^H \tilde{\mathrm{d}}(n)$$
$$= \begin{bmatrix} a_1^H & \cdots & a_L^H & -b^H \end{bmatrix} \begin{bmatrix} v_1(n) e^{-j\theta_1} \\ \vdots \\ v_L(n) e^{-j\theta_L} \\ \tilde{\mathrm{d}}(n) \end{bmatrix} = \mathrm{w}^H \mathbf{u}(n) \tag{10}$$

Where, $\mathrm{w}(n)=[\mathrm{a}(n)\ \text{-}\mathrm{b}(n)]^{\mathrm{T}}$ denotes overall equalizer coefficient vector, $\mathrm{u}(n) = [\mathrm{x}(n)\ \tilde{\mathrm{d}}(n)]^T$ denotes composite input vector, $\mathrm{a}_k^H = [a_0^k \cdots a_{N-1}^k]^*$ denotes the coefficients vector of feedforward filter, $\mathrm{b}^H = [b_1 \cdots b_M]^*$ denotes the coefficients vector of feedback filter, N and M denote the feedforward and feedback filter taps respectively, * denotes complex conjugate, ^{H} denotes transpose conjugate, ^{T} denotes transpose. $\tilde{\mathrm{d}}(n) = \left[\tilde{d}(n-1) \cdots \tilde{d}(n-M)\right]^T$ denotes the vector of M previously detected symbols stored in the feedback filter, $p(n)$ represents the output of the linear part of the equalizer, it can be written as

$$p(n) = \sum_{k=1}^{K} a_k^H v_k(n) e^{-j\theta_k} = \sum_{k=1}^{K} p_k(n) \tag{11}$$

As shown in Fig.6., the baseband signals firstly perform carrier phase compensation. Using DPLL technology (Proakis, 2003), we can obtain carrier phase compensation as follows

$$\hat{\theta}_{k+1}(n) = \theta_k(n) + K_{f_1} \phi_k(n) + K_{f_2} \sum_{m=1}^{n} \phi_k(m) \tag{12}$$

Where, $\Phi_k(n+1) = \mathrm{Im}\left\{p_k(n)\left[p_k(n) + e(n)\right]^*\right\}$, K_{f_1} and K_{f_2} are constants, $K_{f_2} \le K_{f_1}$.

A fast self-optimized LMS (FOLMS) algorithm (Bragard & Jourdain, 1990) is used to update the equalizer vector $\mathrm{w}(n)$. But in (Bragard & Jourdain, 1990), the formulations are conducted based on the single channel line equalizer (LE). In this chapter, we extent it to multi-channel decision feedback equalizer and consider the effect of carrier phase compensation. It can be deducted by the composite input data $\mathrm{u}(n)$ and the error signal $e(n)$. So, we can rewrite the FOLMS algorithm as follows

$$\mathbf{w}(n+1) = \mathbf{w}(n) + \mu(n) \mathbf{x}(n) e^*(n) \tag{13}$$

$$\mu(n+1) = \mu(n) + \beta \mathrm{Re}\left[\mathbf{G}^{\mathrm{H}}(n) \mathbf{x}(n) e^*(n)\right] \tag{14}$$

$$g(n) = \mathbf{x}^{\mathrm{H}}(n) \mathbf{G}(n) \tag{15}$$

$$x'(n) = e^*(n) / \mu(n) \tag{16}$$

$$\xi(n) = x'(n) - g(n) \tag{17}$$

$$\mathbf{G}(n+1) = \mathbf{G}(n) + \mu(n)\mathbf{x}(n)\xi(n) \tag{18}$$

Where, $\mu(n)$ is the step-size factor for controlling the convergence ratio of the equalizer, which can adaptively update, $g(n)$ is temporary variant for updating $\mu(n)$, α is constant. $\mathrm{Re}(\cdot)$ denotes the real part of data.

3.2.2 Decoding of SCTCM

In order to simplify the decoding algorithm of SCTCM, the symbol decoding of SCTCM is transformed into bit decoding through calculating the LLR of coded bits. The position of LLR calculator is shown in Fig.6. The calculation is detailed as follows.

For MPSK, the corresponding $m = \log_2 M$ coded bits are mapped to an M-ary signal. The probability $p(b_i = 1 \mid y_k)$ of ith coded bit of kth received symbol can be calculated as

$$\begin{aligned} p(b_i = 1 \mid y_k) &= \frac{p(y_k \mid b_i = 1) \cdot p(b_i = 1)}{p(y_k)} \\ &= \frac{1}{p(y_k)} \left\{ \sum_{b_1} \sum_{b_2} \cdots \sum_{b_m} p(y_k \mid b_i = 1, b_1, \ldots, b_m) \cdot p(b_i = 1, b_1, \ldots, b_m) \right\} \end{aligned} \tag{19}$$

Let the probabilities $p(b_i = 1)$ and $p(b_i = 0)$ of coded bit b_i are the same. Therefore:

$$p(b_i = 1, b_1, \ldots, b_m) = p(b_1 = 1) \cdot p(b_2) \cdots p(b_m) = \frac{1}{2^m} \tag{20}$$

So, (19) can be simplified as

$$p(b_i = 1 \mid y_k) = \frac{1}{2^m \cdot p(y_k)} \sum_{b_1} \sum_{b_2} \cdots \sum_{b_m} p(y_k \mid b_i = 1, b_1, \ldots, b_m) \tag{21}$$

The probability $p(b_i = 0 \mid y_k)$ of ith coded bit of kth received symbol can be calculated as

$$p(b_i = 0 \mid y_k) = \frac{1}{2^m \cdot p(y_k)} \sum_{b_1} \sum_{b_2} \cdots \sum_{b_m} p(y_k \mid b_i = 0, b_1, \ldots, b_m) \tag{22}$$

The LLR value of ith coded bit of kth received symbol is

$$\Lambda_i = \ln \frac{p(b_i = 1 \mid y_k)}{p(b_i = 0 \mid y_k)} = \ln \frac{\sum_{b_1} \sum_{b_2} \cdots \sum_{b_m} p(y_k \mid b_i = 1, b_1, \ldots, b_m)}{\sum_{b_1} \sum_{b_2} \cdots \sum_{b_m} p(y_k \mid b_i = 0, b_1, \ldots, b_m)} \tag{23}$$

From the received signal $y_k = d_k + n_k$,and from the noise distribution it follows that $p(y_k \mid d_k)$ is given by

$$p(y_k \mid d_k) = \exp\left(-(y_k - d_k)^2 / (2\sigma^2)\right) \Big/ \sqrt{2\pi\sigma^2} \tag{24}$$

So, the (23) can be calculated as

$$\Lambda_{k,i} = \ln \frac{\sum_{d \in \mathrm{B}_{i=1}} \exp\left(-\frac{(\hat{d}_k - d)^2}{2\sigma^2} \right)}{\sum_{d \in \mathrm{B}_{i=0}} \exp\left(-\frac{(\hat{d}_k - d)^2}{2\sigma^2} \right)} \tag{25}$$

where, $\mathrm{B} = \{d_1, d_2, \cdots, d_M\}$ denotes the finite alphabet used for MPSK signals, $\mathrm{B}_{i=1}$ and $B_{i=0}$ denote the sets of all possible symbol values, in which the *i*th coded bit is 1 and 0 respectively.

We can simplify symbol decoding into bit decoding using Eq.(25). The probability distributions of the output sequences $\tilde{P}_k^O(u)$ and $\tilde{P}_k^O(c)$ can be calculated as follows:

$$\tilde{P}_k^O(u) = \tilde{B}_u \sum_{e:u(e)=u} \alpha_{k-1}\left[s^S(e)\right] P_k^I\left[u(e)\right] P_k^I\left[c(e)\right] \beta_k\left[s^E(e)\right] \tag{26}$$

$$\tilde{P}_k^O(c) = \tilde{B}_c \sum_{e:c(e)=c} \alpha_{k-1}\left[s^S(e)\right] P_k^I\left[u(e)\right] P_k^I\left[c(e)\right] \beta_k\left[s^E(e)\right] \tag{27}$$

where, $\tilde{B}_u$ and $\tilde{B}_c$ are normalization constants as

$$\tilde{B}_u \to \sum_u P_k^O(u) = 1 \tag{28}$$

$$\tilde{B}_c \to \sum_c P_k^O(c) = 1 \tag{29}$$

The forward recursion $\alpha_k[\cdot]$ and backward recursion $\beta_k[\cdot]$ are given by

$$\alpha_k(s) = \sum_{e:s^E(e)=s} \alpha_{k-1}\left[s^S(e)\right] P_k^I\left[u(e)\right] P_k^I\left[c(e)\right] \ , \ k = 1, 2, \ldots, n \tag{30}$$

$$\beta_k(s) = \sum_{e:s^S(e)=s} \beta_{k+1}\left[s^S(e)\right] P_{k+1}^I\left[u(e)\right] P_{k+1}^I\left[c(e)\right] \ , \ k = n-1, \ldots, 0 \tag{31}$$

The $\alpha_k[\cdot]$ computation will be initialized as

$$\alpha_0(s) = \begin{cases} 1, & s = S_0 \\ 0, & other \end{cases} \tag{32}$$

If the trellis is terminated to a known state S_N, then the $\beta_k[\cdot]$ computation will be initialized as

$$\beta_n(s) = \begin{cases} 1, s = S_N \\ 0, other \end{cases} \tag{33}$$

Otherwise

$$\beta_n(s) = 1 / M_s \ , \ \forall s \tag{34}$$

In this chapter, log-map algorithm (Soleymani & Gao, 2002) is used to simplify calculation through transforming multiplication into addition.

3.2.3 Soft symbol estimation

As shown in Fig.6, the LLR values of coded bits, output from decoder of SCTCM, are used to implement symbol estimation. And then, these symbols are fed back to feedback filter of MC-ADFE to perform joint iterative scheme. So, the symbol estimation is key module to perform soft IED.

There are two methods to estimate data symbol: hard estimation and soft estimation. Compare with hard estimation, soft estimation can void error symbols spread during the course of iterations. What's more, soft estimation can more sufficiently utilize decoding gain to update system performance. In this chapter, soft method is adopted to estimate data symbols.

The soft symbol estimation can be obtained as follows:

$$\hat{d}_k = \sum_{d \in \mathrm{B}} d \prod_{i=1}^{m} p(b_i) \tag{35}$$

where, $\mathrm{B} = \{d_1, d_2, \cdots, d_M\}$ denotes the finite alphabet used for MPSK, b_i denotes the ith coded bit, $i = 1, 2, \cdots m$.

The probability distributions $P(b_i)$ of coded bits can be obtained from the corresponding LLR values $L(b_i)$. Therefore:

$$P(b_i) = \frac{e^{b_i \cdot L(b_i)}}{1 + e^{L(b_i)}} \tag{36}$$

4. Simulation results

In this section, we use simulation experiments to verify the performance of the proposed soft IED algorithm. The system parameters of computer simulation are shown in Table 1. The outer decoders of SCTCM adopt convolutional codes encoder with 4-state 1/2 code rate for QPSK modulation. In this paper, we integrate the multi-path fading and additive white Gaussian noise to simulate underwater acoustic channel.

Based on the sound speed profile (SSP) measured in the lake and finite-element ray (FER) tracing method (Bellhop) (Porter & Liu, 1994), the channel impulse response is shown in

Fig.7. From the SSP (shown in Fig.1.), we know that the water with mixed gradient SSP is about 53m deep. The transmission distance is 2000m.

Carrier frequency	10 KHz
Symbol rate	5 Kbps
Doppler	10 Hz
Array elements	4
Training symbols	200 sys

Table 1. Simulation parameters

4.1 Soft iterative performance of iterative equalization and decoding

The system parameters of simulation are shown in Table 1. The channel impulse responses are shown in Fig.7. As shown in Fig.7, the multi-path propagation is very seriously.

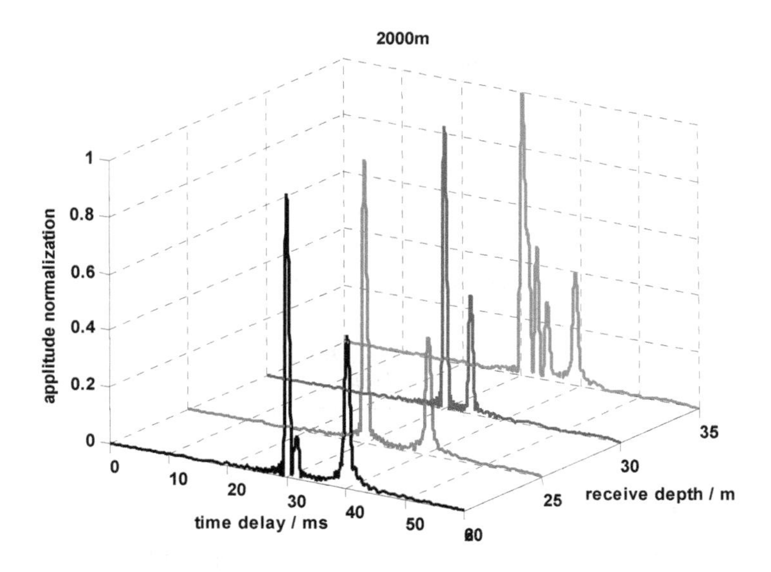

Fig. 7. Channel impulse response (2000m)

Fig.8 shows the BER curves of IED algorithm with soft iteration for QPSK modulation. As show in Fig.8, the iteration algorithm can sufficient utilize the decoding gain provided by decoder of SCTCM to enhance the equalizer performance such that the system performance is increased and the data transmission with lower BER can be obtained.

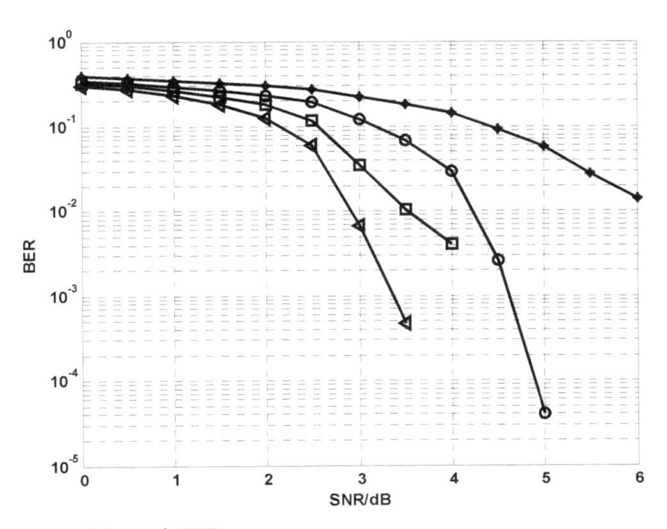

Fig. 8. Performance of the soft IED

4.2 Dual model of iterative equalization and decoding

The dual model IED is shown in Fig.9. There are two parts: (1) iterative equalization and decoding; (2) iterative decoding. As mentioned in Section 3, the mutichannel ADFE and decoder of SCTCM exchange soft information in an iterative manner in the IED scheme. So, we can perform decoding iteration before IED. And thus, the accuracy of symbol estimation is further improved such that the equalizer performance is improved greatly.

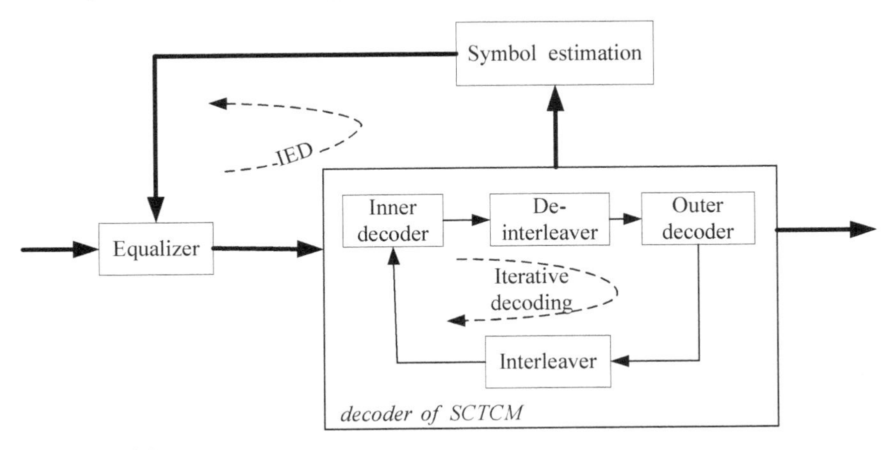

Fig. 9. Dual model IED

As shown in Table 2, in dual model IED, the soft symbols estimation, based on the coded bits output from decoder of SCTCM, are more accurated. And thus, the propagation errors can be futherly reduced.

IED iteration	0		1		2	
Decoder iteration	0	1	0	1	0	1
3 dB	0.2217	0.2072	0.1205	0.0824	0.0351	0.0185
3.5 dB	0.1856	0.1582	0.0682	0.0326	0.0105	0.0042
4 dB	0.1451	0.0987	0.0295	0.0123	0.0041	0.0024
4.5 dB	0.0941	0.0358	0.0026	1e-4	0	0
5 dB	0.0589	0.0097	4e-5	0	0	0
5.5 dB	0.0282	0.0016	0	0	0	0
6 dB	0.0141	3e-4	0	0	0	0

Table 2. Performance of dual model IED

5. Conclusions

In this chapter, according to the characteristics of underwater acoustic channel, SCTCM technology with rate-1 inner code, is adopted to improve the bandwidth efficiency of underwater acoustic channel. Simultaneously, LLR calculation is introduced to simplify symbol decoder into bits decoder. The soft IED scheme with soft symbol estimation is proposed to overcome the multi-path fading of underwater acoustic channel and enhance the performance of equalizer through utilizing decoding gain provided by decoder and the information symbols with soft symbol estimation fed back to equalizer so that the performance of communication system is improved greatly. What's more, the dual model IED scheme is proposed to obtain lower BER in order to meet the demands of higher system performance. The simulation results verify the proposed algorithm can obtain satisfied data transmission with the small iterations, especially the dual model IED.

6. References

Bragard P., Jourdain G.. A fast self-optimized algorithm for non-stationary identification: application to underwater equalization, *Proceedings of IEEE ICASSP*, pp. 1425-1428, 1990.

Berrou C., Glavieux A., and Thitimajshima P.. Near Shannon Limit Error-Correcting Coding:Turbo Codes, *Proceeding of IEEE Int. Conf on Communications*, pp. 1064-1070, Geneva, Switzerland, May 1993.

Benedetto S., Montorsi G.. Serial concatenation interleaved codes: performance analysis, design and iterative decoding. TDA Progress Report 42-126, 1996.

Benedetto S., Divsalar D., Montorsi G., Pollara F.. Parrallel concatenated trellis coded modulation, *Proceeding of IEEE ICC'96*, Vol.2, pp. 974-978.

Benedetto S., Divsalar D., Montorsi G., Pollara F.. Serial Concatenated Trellis Coded Modulation with Iterative Decoding: Design and Performance, *IEEE Global Telecommunications Conference*, (CTMC), Nov. 1997.

Benedetto S., Divsalar D., Montorsi G., Pollara F.. Serial concatenation of interleaved codes: performanceanalysis, design, and iterative decoding. *IEEE Trans. on Information Theory*, Vol. 44, May 1998, pp. 909-926.

Berthet, A.O.; Visoz, R.; Unal, B.; Tortelier, P.. A comparison of several strategies for iteratively decoding serially concatenated convolutional codes in multipath Rayleigh fading environment, *Proceeding of Global Telecommunications Conference*, Vol.2, pp. 783-789, 2000.

Chung S.Y., Lou H.. TCM and turbo TCM for audio broadcasting, *Proceeding of Vehicular Technology Conference*, Vol.3, pp. 2222-2226, Tokyo, Japan, May 15-18, 2000.

Choi J.W.. Iterative Multi-Channel Equalization and Decoding for High Frequency Underwater Acoustic Communications, *IEEE SAM*, 2008.

Divsalar D., Pollara F.. Hybrid Concatenated Codes and Iterative Decoding, JPL TMO Prog. Report, Aug. 15, 1997.

Divsalar D., Dolinar S., Pollara F.. Serial concatenated trellis coded modulation with rate-1 inner code, *Proceedings of IEEE Global Telecommunications Conference*, Vol.2, pp. 777-782, Nov.27-Dec.1, 2000.

Flanagan M.F., Fagan A.D.. Iterative Channel Estimation, Equalization, and Decoding for Pilot-Symbol Assisted Modulation Over Frequency Selective Fast Fading Channels. *IEEE Transactions on Vehicular Technology*, Vol. 56, No. 4, July 2007, pp. 1661-1670.

Ho M. "Performance bounds for serially-concatenated trellis-coded modulation", Signals, Systems & Computers, 1997. Conference Record of the Thirty-First Asilomar Conference on Volume 2, 2-5 Nov. 1997:1364 - 1368.

Kilfoyle D.B. and Baggeroer A.B.. The state of the art in the underwater acoustic communications. *IEEE Journal of Oceanic Engineering*, Vol. 25, No. 1, 2000, pp. 4-27.

Koetter R., Singer A.C., Tuchler M.. Turbo equalization. *IEEE Signal Processing Magazine*, Vol. 21, No. 1, Jan. 2004, pp. 67-80.

Legoff S., Glavieux A., and Berrou C.. Turbo codes and high spectral efficiency modulation, *Proceeding of IEEE ICC'94*, New Orleans, LA, May 1-5 1994.

Li X., Song W.T., Luo H.W.. Joint turbo equalization and turbo TCM for mobile communication systems, *Proceeding of 12th IEEE International Symposium on Personal, Indoor and Mobile Radio Communications*, Vol.1, pp. A184-A188, Sep. 30.-Oct. 3, 2001.

Noorbakhsh M., Mohamed-Pour K.. Combined turbo equalisation and block turbo coded modulation, *Proceeding of IEE Communications*, Vol. 150, No. 3, pp. 149-152, June 2003.

Otnes, R., Tuchler M.. Iterative channel estimation for turbo equalization of time-varying frequency-selective channels. *IEEE Transactions on Wireless Communications*, No.6, Nov. 2004, pp. 1918-1923.

Porter M.B., Liu Y.C.. Finite-Element Ray Tracking. *Theoretical and computational Acoustics*, Vol. 2, 1994, pp. 947-956.

Proakis J.G.. Digital communication (4th Edition) . Beijing: Publishing House of Electronics Induxtry, 2003.

Robertson P., Woerz T.. Bandwidth efficient turbo trellis coded modulation using punctured component codes. *IEEE Journal on Selected Areas in Communications*, Vol.16, No.2, Feb.1998.

Stojanovic M., Catipovic J., Proakis J.. Recent advances in high-speed underwater acoustic communications. *IEEE J .Oceanic Eng.*, Vol. 21, No. 2, April 1996, pp. 125-136.

Soleymani M.R., Gao Y.Z., Vilaipornsawai U.. Turbo Coding for Satellite and Wireless CommunicationsBoston. MA: Kluwer Academic Publishers, 2002.

Shohon T., Jinbo K., Ogiwara H. Performance improvement with increasing interleaver length for serially concatenated trellis coded modulation. *Proceeding of IEEE International Symposium on Information Theory*, June 29-July 4, 2003.

Stojanovic M., Frieitag L.. Wideband Underwater CDMA: Adaptive Multichannel Receiver design, *Proceeding of MTS Oceans 2005*, pp. 1-6, USA 2005.

Tuchler M., Singer A.C., Koetter R.. Minimum mean squared error equalization using a priori information. *IEEE Transactions on Signal Processing*, Vol.50, No.3, March 2002, pp. 673-683.

Ungerboeck G.. Channel coding with multilevel phase signaling. *IEEE Trans. Inf. Th.*, Vol.IT-25, Jan. 1982, pp. 55-67.

Yang P., Ge J.H.. Combination of turbo equalization and turbo TCM for mobile communication system, *Proceeding of IEEE International Symposium on Communications and Information Technology*, Vol.1, pp. 383-386, Oct.12-14, 2005.

Zhao L., Zhu W.Q., Zhu M.. Adaptive Equalization Algorithms for Underwater Acoustic Coherent Communication System. *Journal of Electronics & Information Technology*, Vol. 30, No. 3, 2008, pp. 648-651.

7

CI/OFDM Underwater Acoustic Communication System

Fang Xu and Ru Xu
Xiamen University
China

1. Introduction

The underwater acoustic channel (UAC) is one of the most challenging environments to be encountered for the communication. Because of the absorption of the signal, the path loss depends on the signal frequency (Berkhovskikh & Lysanov, 2003; Jensen et al., 2011). Multipath transmission causes intersymbol interference (ISI), and it extends over tens to hundreds of milliseconds according to the communication distance (Stojanovic & Preisig, 2009). Since the velocity of sound in water is about 1500m/s, any relative motion includes the transmitter or receiver and even surface waves will cause non-negligible Doppler effects, including shifting and spreading. All these phenomena dramatically limit the data rate achievable and the performance of the communications. The bandwidth is very limited, and the system is actually a broadband communication system because the center frequency of the signal is always at the same order of the bandwidth (Stojanovic, 1996; Stojanovic, 2007; Stojanovic & Preisig, 2009).

In order to achieve high data rate, it is important to use bandwidth-efficient modulation methods in UAC. Multi-carrier modulation is one of the candidates that can be used. Orthogonal Frequency Division Multiplexing (OFDM) (Lam & Ormondroyd, 1997; Kim & Lu, 2000; Stojanovic, 2006; Stojanovic, 2008; Li et al., 2008), direct-sequence spread-spectrum (DSSS) (Freitag et al. 2001; Frassati et al. 2005), frequency-hopped spread-spectrum (FHSS) (Stojanovic, 1998; Freitag et al., 2001) and code-division multiple access (CDMA) (Charalampos et al. 2001; Stojanovic & Freitag, 2006; Tsimenidis, 2001)were used in UAC channels in recent years and much literature focus on the conceptual system analysis and computer simulations.

In this chapter, we introduce a new multi-carrier modulation into the UAC channels which is called Carrier Interferometry OFDM (CI/OFDM) (Nassar et al., 1999; Wiegandt & Nassar, 2001; Nassar et al., 2002a, 2002b). Compared with OFDM, the CI/OFDM has a low PAPR characteristic and inherent frequency selective combining, which makes it a very attractive signaling scheme in frequency selective fading channels (Wiegandt & Nassar, 2001; Wiegandt et al., 2001; Wiegandt & Nassar, 2003; Wiegandt et al., 2004).

The chapter is organized as follows. In Section II, the characteristics of CI signal are analyzed. Two algorithms are proposed in Section III. Details are focused on the PAPR performance, and new algorithms to complete the modulation and demodulation of the

CI/OFDM. The configuration of the CI/OFDM underwater acoustic communication system is presented in Section IV. Furthermore, the key algorithms including synchronization, channel estimation and equalization are described. In Section V, Performance results for different field tests are summarized. Conclusions are drawn in Section VI.

2. CI/OFDM signals

2.1 The theory of the CI/OFDM

In CI/OFDM transmitter, after serial to parallel transform, information symbols are modulated onto all the N parallel subcarriers and then added linearly together to get the output signal (Nassar et al. 2002). As shown in Fig.1 , the output of the signal is

$$s(t) = \sum_{k=0}^{N-1} s_k(t) = \mathrm{Re}\left\{ \sum_{k=0}^{N-1} a_k c_k(t) \right\} \quad 0 \le t \le T \tag{1}$$

where $s_k(t)$ is the modulated signal for the kth information symbol a_k. $\mathrm{Re}(\cdot)$ is the real part of the signal and $c_k(t)$ is the kth CI signal, which can be express by

$$c_k(t) = \sum_{i=0}^{N-1} e^{j2\pi i \Delta f t} \cdot e^{j\theta_k^i} \tag{2}$$

It is easy to see that $c_k(t)$ is a multi-carrier signal with different phase offsets $\theta_k^i = (2\pi/N) \cdot k \cdot i$. Submit (2) into (1), we get the continuous baseband transmitted signal

$$s(t) = \mathrm{Re}\left\{ \sum_{k=0}^{N-1} a_k \sum_{i=0}^{N-1} e^{j2\pi i \Delta f t} \cdot e^{j\frac{2\pi}{N}ki} \right\} \quad 0 \le t \le T \tag{3}$$

We rewrite the discrete form of (3) with the Nyquist sampling rate of $f_s = N\Delta f$

$$s(n) = \mathrm{Re}\left\{ \sum_{k=0}^{N-1} a_k \sum_{i=0}^{N-1} e^{j\frac{2\pi}{N}in} \cdot e^{j\frac{2\pi}{N}ik} \right\} \quad n = 0,1,\ldots,N-1 \tag{4}$$

where $\Delta f = 1/T_s$ (T_s is one CI/OFDM symbol duration) to ensure orthogonality among subcarriers, and $(2\pi/N)k \cdot i$ is the phase offset used for a_k which ensures the orthogonality among the N information symbols.

After transmitted over a frequency selective fading channel, the received signal at receiver side is

$$r(n) = \sum_{k=0}^{N-1} \sum_{i=0}^{N-1} \alpha_i a_k e^{j\frac{2\pi}{N}in} \cdot e^{j\frac{2\pi}{N}ik} \cdot e^{j\phi_i} + w(n) \quad n = 0,1,\ldots,N-1 \tag{5}$$

where α_i and ϕ_i are the amplitude fade and phase offset on the ith carrier, respectively. $w(n)$ is the addictive white Gaussian noise (AWGN).

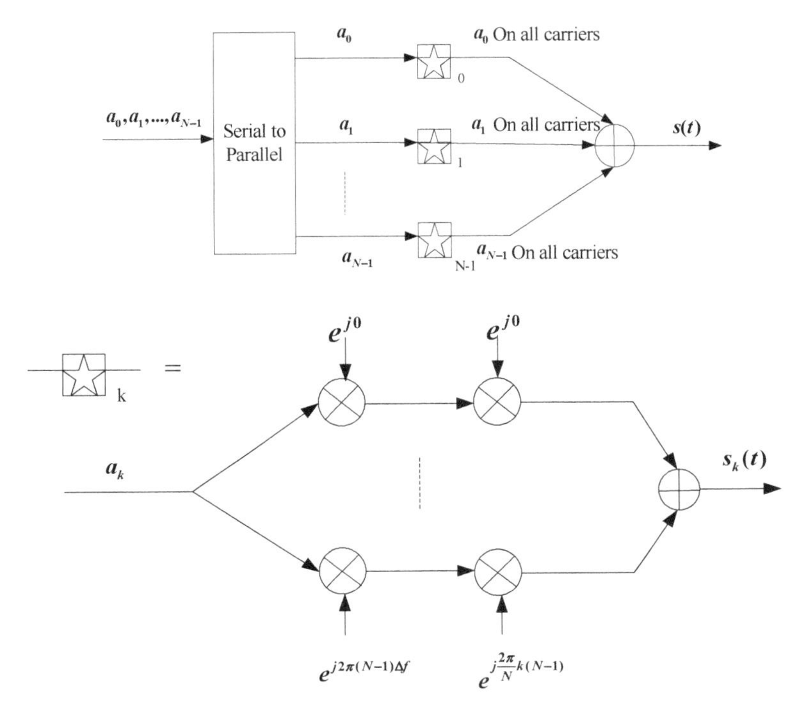

Fig. 1. The conceptual CI/OFDM Transmitter

Fig.2 depicts the modulation theory of CI/OFDM in transmitter and the detection of the kth symbol signal at the receiver side (Nassar et al. 2002). Assume perfect synchronization, the received signal is first projected onto the N orthogonal carriers, multi-carrier demodulation and phase offsets remove are carried out after that. This leads to the decision vector $r^k = (r_0^k, r_1^k, \ldots, r_{N-1}^k)$ for the information symbol a_k, where r_i^k is defined as

$$r_i^k = A\alpha_i a_k + A \cdot \sum_{j=0, j\neq k}^{N-1} \alpha_i a_j \cos(\theta_k^i - \theta_j^i) + w_i \tag{6}$$

The first part of (6) is the desired information symbol which is randomly faded by factor α_i, and the second part is the interferences of the other N-1 information symbols which modulated on the same carrier.

Different combining strategies are employed to help restore orthogonality between subcarriers. In AWGN channel, the optimal combining is equal gain combining (EGC). After performing $C = \sum_{i=0}^{N-1} r_i$, interferences are close to zero. While in frequency selective channel, different combining strategies are used to get combining gains, for example, the maximum ratio combining (MRC), the minimum mean square error combining (MMSEC) (Itagkai & Adachi, 2004). After combining, the signals are sent to the detector.

As presented to date, the implementation of original CI/OFDM is complicated, and it is important to note that the receiver is designed for detecting only one information symbol. Although CI/OFDM had been proved that it could improve BER performance by exploiting frequency diversity and depress the PAPR simultaneously, its implementation was complicated and only conceptual transmitter and receiver models had been given in the literature.

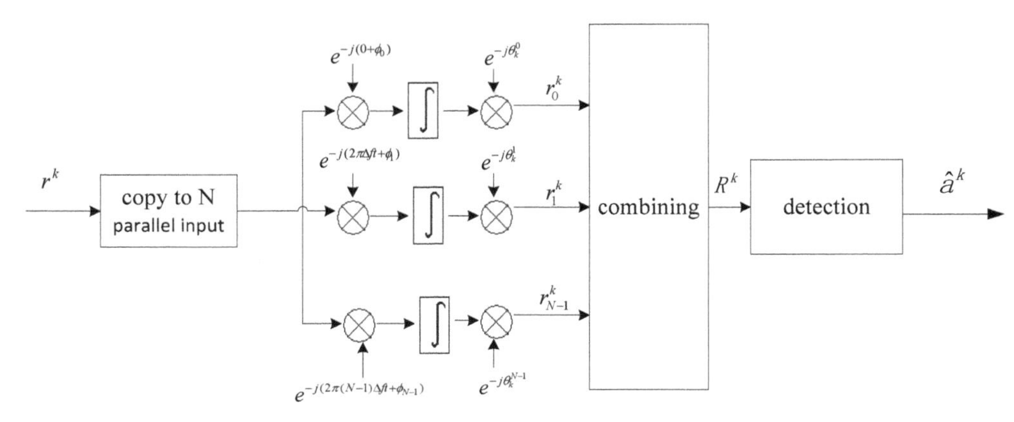

Fig. 2. The CI/OFDM receiver for the kth symbol

2.2 The characteristics of the CI signal

The baseband CI signal (Nassar et al. 2002) is given bellow

$$c(t) = \sum_{i=0}^{N-1} e^{j2\pi i \Delta f t} \tag{7}$$

where N is the number of the subcarriers and Δf is the interval of the subcarriers. It is obviously that CI signal is a periodic signal. Simulation results are shown in Fig. 3. In simulations, $N = 8$, $\Delta f = 1Hz$, the width of the main lobe and the side lobe are $2/(N\Delta f) = 0.125s$ and $1/(N\Delta f) = 0.0625s$, respectively. By selecting optimal phase offsets, CI signals are orthogonal to each other.

We rewrite the discrete CI signal as bellow

$$c(n) = \sum_{i=0}^{N-1} e^{j\frac{2\pi}{N}in} \tag{8}$$

It is clearly that the discrete CI signal is the result of sampling a rectangle pulse with the sampling rate Δf in the frequency domain. It has constant amplitude (CA).

Based on the analysis of the CI signal, two novel algorithms for CI/OFDM modulation and demodulation are presented in this chapter.

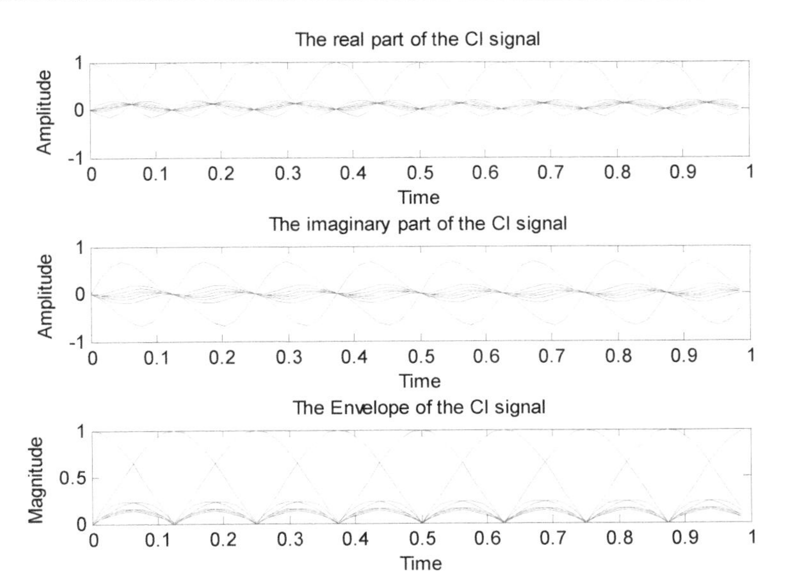

Fig. 3. Baseband CI signals

3. Proposed algorithms

3.1 Multi-carrier algorithm

As in (4), the discrete kth transmitted symbol is

$$s(n)=\sum_{i=0}^{N-1}(a_k\cdot e^{j\frac{2\pi}{N}ik})e^{j\frac{2\pi}{N}in}\qquad n=0,1,\ldots,N-1 \tag{9}$$

It is obviously that (9) is an inverse discrete Fourier transform (IDFT) with weighting efficients $a_k\cdot e^{j\frac{2\pi}{N}ik}$ which change with the index k and i. The IDFT weighting coefficient can be written as a matrix

$$\begin{bmatrix} a_0\cdot e^{j\frac{2\pi}{N}i\cdot 0} \\ a_1\cdot e^{j\frac{2\pi}{N}i\cdot 1} \\ \vdots \\ a_k\cdot e^{j\frac{2\pi}{N}i\cdot k} \\ \vdots \\ a_{N-1}\cdot e^{j\frac{2\pi}{N}i\cdot(N-1)} \end{bmatrix} = \begin{bmatrix} a_0e^{j\frac{2\pi}{N}\cdot 0\cdot 0} & a_0e^{j\frac{2\pi}{N}\cdot 1\cdot 0} & \cdots & a_0e^{j\frac{2\pi}{N}\cdot i\cdot 0} & \cdots & a_0e^{j\frac{2\pi}{N}\cdot(N-1)\cdot 0} \\ a_1e^{j\frac{2\pi}{N}\cdot 0\cdot 1} & a_1e^{j\frac{2\pi}{N}\cdot 1\cdot 1} & \cdots & a_1e^{j\frac{2\pi}{N}\cdot i\cdot 1} & \cdots & a_1e^{j\frac{2\pi}{N}\cdot(N-1)\cdot 1} \\ \vdots & \vdots & \vdots & \vdots & \vdots & \vdots \\ a_ke^{j\frac{2\pi}{N}\cdot 0\cdot k} & a_ke^{j\frac{2\pi}{N}\cdot 1\cdot k} & \vdots & a_ke^{j\frac{2\pi}{N}\cdot i\cdot k} & \vdots & a_ke^{j\frac{2\pi}{N}\cdot(N-1)\cdot k} \\ \vdots & \vdots & \cdots & \vdots & \cdots & \vdots \\ \underbrace{a_{N-1}e^{j\frac{2\pi}{N}\cdot 0\cdot(N-1)}}_{0^{\#}\text{subcarrier}} & \underbrace{a_{N-1}e^{j\frac{2\pi}{N}\cdot 1\cdot(N-1)}}_{1^{\#}\text{subcarrier}} & \cdots & \underbrace{a_{N-1}e^{j\frac{2\pi}{N}\cdot i\cdot(N-1)}}_{i^{\#}\text{subcarrier}} & \cdots & \underbrace{a_{N-1}e^{j\frac{2\pi}{N}\cdot(N-1)\cdot(N-1)}}_{(N-1)^{\#}\text{subcarrier}} \end{bmatrix} \tag{10}$$

Since i is the index of subcarrier, the columns of the matrix are corresponding to different subcarriers, that is

$$\left[\underbrace{\sum_{k=0}^{N-1} a_k e^{j\frac{2\pi}{N}\cdot 0\cdot k}}_{0^{\#}\text{ subcarrier}} \quad \underbrace{\sum_{k=0}^{N-1} a_k e^{j\frac{2\pi}{N}\cdot 1\cdot k}}_{1^{\#}\text{ subcarrier}} \quad \cdots \quad \underbrace{\sum_{k=0}^{N-1} a_k e^{j\frac{2\pi}{N}\cdot i\cdot k}}_{i^{\#}\text{subcarrier}} \quad \cdots \quad \underbrace{\sum_{k=0}^{N-1} a_k e^{j\frac{2\pi}{N}\cdot(N-1)\cdot k}}_{(N-1)^{\#}\text{ subcarrier}}\right] \tag{11}$$

(11) implies that the coefficient of the IDFT can also be corresponding to another IDFT (Xu et al., 2007a, 2007b). Hence, the CI/OFDM modulation model employed in this chapter corresponds to

$$\begin{aligned} s(n) &= \sum_{k=0}^{N-1} a_k c_k\left(\frac{nTs}{N}\right) \\ &= \sum_{k=0}^{N-1} a_k \sum_{i=0}^{N-1} e^{j\frac{2\pi}{N}ki} e^{j\frac{2\pi}{N}ni} \\ &= \sum_{i=0}^{N-1} \left(\sum_{k=0}^{N-1} a_k e^{j\frac{2\pi}{N}ki}\right) e^{j\frac{2\pi}{N}ni} \\ &= N \cdot \underset{n}{IDFT}\left[N \cdot \underset{i}{IDFT}(a_k)\right] \\ &= N^2 \cdot \underset{n}{IDFT}\left[\underset{i}{IDFT}(a_k)\right] \qquad n = 0,1,\ldots,N-1 \end{aligned} \tag{12}$$

Fig.4 shows a block diagram of the proposed system. At the transmitter side, the input data is first mapped into a baseband constellation. Then the data sequence is converted to parallel and enters the first IDFT to perform CI spreading. After that, the second IDFT is

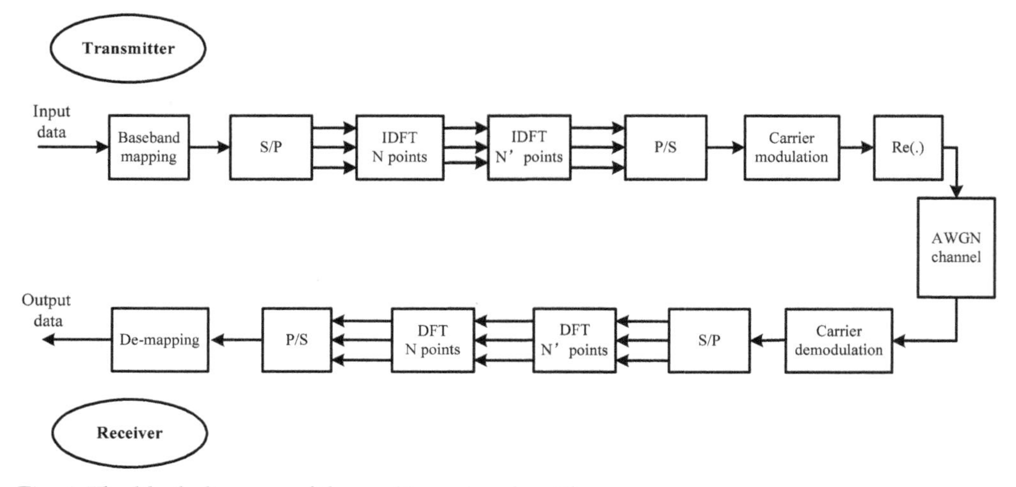

Fig. 4. The block diagram of the multi-carrier algorithm

used to implement orthogonal multi-carrier modulation. Parallel data is first transformed to serial data, then the complex base-band signal is then up-converted to the transmission frequency, and the real part of the signal is sent out to the channel. In the receiver, the signal is first down-converted to the base-band. Serial to parallel transformation followed by orthogonal multi-carrier demodulated which completed by the first discrete Fourier transform (DFT). Then CI code de-spreading is implemented by the second DFT and finally, the phase constellation of the data is extracted (Xu et al. 2007).

3.2 Single-carrier algorithm

The CI spread code in (10) is similar to the polyphase codes (Heimiller, 1961). Polyphase codes were proven to have good periodic correlation properties, the sequence is

$$1,1,...,1,1,\xi_1,\xi_2,\xi_3,....,\xi_{p-1},1,\xi_1^2,\xi_2^2,...,\xi_{p-1}^2,1,\xi_1^3,\xi_2^3,...,\xi_{p-1}^3,1,\xi_1^{p-1},\xi_2^{p-1},...,\xi_{p-1}^{p-1} \quad (13)$$

where $\xi_k = e^{-j(2\pi k/p)}$, $0 \le k \le p-1$ is a primitive Pth root of unity, the sequence has zero periodic correlation except for the peaks at $i = 0, p^2, 2p^2, ...$.

Polyphase code were proven to be a constant amplitude , zero autocorrelation (CAZAC) sequence (Heimiller, 1961). According to the characteristics of the CAZAC sequence, if u_i is a CAZAC sequence, then $\overline{u}_i$, where $\overline{u}$ denotes complex conjugation, is also a CAZAC sequence (Milewski, 1983). Note that the orthogonality, periodicity, constant amplitude and zero autocorrelation are not changed, it suggests a new way of thinking about constructing new CI signals.

In this chapter, the new CI signals are complex conjugations of primary CI signals (Nassar et al. 2002), which can be written as

$$\begin{bmatrix} a_0 \cdot e^{-j\frac{2\pi}{N}i\cdot 0} \\ a_1 \cdot e^{-j\frac{2\pi}{N}i\cdot 1} \\ \vdots \\ a_k \cdot e^{-j\frac{2\pi}{N}i\cdot k} \\ \vdots \\ a_{N-1} \cdot e^{-j\frac{2\pi}{N}i\cdot(N-1)} \end{bmatrix} = \begin{bmatrix} \underbrace{\begin{matrix} a_0 e^{-j\frac{2\pi}{N}\cdot 0\cdot 0} \\ a_1 e^{-j\frac{2\pi}{N}\cdot 0\cdot 1} \\ \vdots \\ a_k e^{-j\frac{2\pi}{N}\cdot 0\cdot k} \\ \vdots \\ a_{N-1} e^{-j\frac{2\pi}{N}\cdot 0\cdot(N-1)} \end{matrix}}_{0^{\#}\text{subcarrier}} & \underbrace{\begin{matrix} a_0 e^{-j\frac{2\pi}{N}\cdot 1\cdot 0} \\ a_1 e^{-j\frac{2\pi}{N}\cdot 1\cdot 1} \\ \vdots \\ a_k e^{-j\frac{2\pi}{N}\cdot 1\cdot k} \\ \vdots \\ a_{N-1} e^{-j\frac{2\pi}{N}\cdot 1\cdot(N-1)} \end{matrix}}_{1^{\#}\text{subcarrier}} & \cdots & \underbrace{\begin{matrix} a_0 e^{-j\frac{2\pi}{N}\cdot i\cdot 0} \\ a_1 e^{-j\frac{2\pi}{N}\cdot i\cdot 1} \\ \vdots \\ a_k e^{-j\frac{2\pi}{N}\cdot i\cdot k} \\ \vdots \\ a_{N-1} e^{-j\frac{2\pi}{N}\cdot i\cdot(N-1)} \end{matrix}}_{i^{\#}\text{subcarrier}} & \cdots & \underbrace{\begin{matrix} a_0 e^{-j\frac{2\pi}{N}\cdot(N-1)\cdot 0} \\ a_1 e^{-j\frac{2\pi}{N}\cdot(N-1)\cdot 1} \\ \vdots \\ a_k e^{-j\frac{2\pi}{N}\cdot(N-1)\cdot k} \\ \vdots \\ a_{N-1} e^{-j\frac{2\pi}{N}\cdot(N-1)\cdot(N-1)} \end{matrix}}_{(N-1)^{\#}\text{subcarrier}} \end{bmatrix} \quad (14)$$

Then, the CI/OFDM signal is mathematically characterized by the following equation

Fig.5 shows the proposed system using the single-carrier algorithm. In the transmitter, the input data is mapping into a baseband constellation. Then the data sequence is converted to parallel and enters the first DFT to perform CI spreading. After that, the IDFT is used to implement orthogonal multi-carrier modulation. The complex baseband signal is then up-converted to the transmission frequency.

$$\begin{aligned} s(n) &= \sum_{k=0}^{N-1} a_k c_k (\frac{nTs}{N}) = \sum_{k=0}^{N-1}\sum_{i=0}^{N-1} a_k e^{-j\frac{2\pi}{N}ki} e^{j\frac{2\pi}{N}ni} \\ &= \sum_{i=0}^{N-1} (\sum_{k=0}^{N-1} a_k e^{-j\frac{2\pi}{N}ki}) e^{j\frac{2\pi}{N}ni} \\ &= N \cdot \underset{n}{IDFT}\left[\underset{i}{DFT}(a_k)\right] \quad n = 0,1,\ldots,N-1 \end{aligned} \tag{15}$$

In the receiver, the signal is down-converted to the baseband. After serial to parallel transformation, the signal is first demodulated by DFT, Then CI de-spreading is implemented by the IDFT and the phase constellation of the data is extracted.

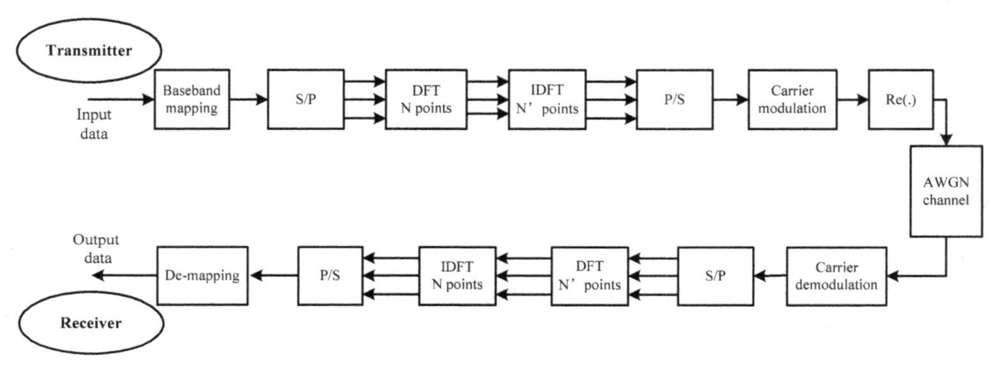

Fig. 5. The block diagram of the single-carrier algorithm

3.3 The comparisons between the conceptual CI/OFDM and the proposed algorithms

In the conceptual model of CI/OFDM, the computational complexity increases with the number of parallel information symbols dramatically, which make it unpractical to the engineering. In the conceptual model of CI/OFDM, the CI spreading needs $N \times N$ complex multiplications $N \times (N-1)$ complex additions. While in multi-carrier algorithm or single-carrier algorithm, only $N/2\log_2 N$ complex multiplications and $N\log_2 N$ complex additions are needed. For example, when $N = 1024$, we need 1048576 complex multiplications and 1046529 complex additions in conceptual CI/OFDM, while only 5120 complex multiplications and 10240 complex additions are needed in our algorithms.

Of course, the two algorithms have their own problems. As in the multi-carrier method, the physical concept is not very clear, since there are two cascaded IDFT in the transmitter which may cause confusion about the transformation between the frequency domain and time domain. On the other hand, in the single-carrier method, filter should be well designed to compress the bandwidth of the output signal.

As shown in fig.6, the performance of CI/OFDM system is verified under AWGN channel. We replace the IDFT by the inverse fast Fourier transform (IFFT) due to the efficiency of the algorithm. It is obviously that there is no difference between the conceptual CI/OFDM and the two algorithms proposed in this chapter.

Since lower PAPR is the most important characteristic of CI/OFDM, and the algorithms presented here are somewhat different from the theoretical realization of the CI/OFDM, it is reasonable for us to verify the PAPR performance based on these two algorithms. Fig. 7 shows the simulation result. A conclusion can be drawn that the two algorithms presented in this chapter have the same PAPR and BER performance as the conceptual CI/OFDM, and lower complexity which make it applicable to engineering.

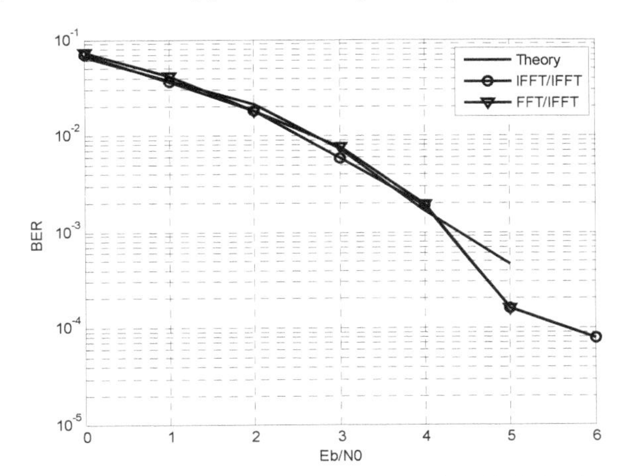

Fig. 6. Performance comparisons of three algorithms under AWGN channels

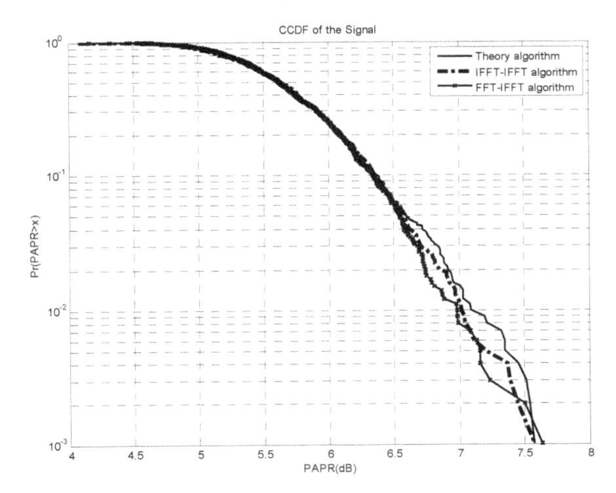

Fig. 7. PAPR Performance comparisons of three algorithms

4. Configuration of the CI/OFDM underwater acoustic communication system

Based on the aforementioned algorithms, two CI/OFDM underwater acoustic communication systems are proposed. Simplified block diagrams of the proposed systems are shown in Fig.8 and Fig.9. We also replace the IDFT by the IFFT due to the efficiency of the algorithm.

As shown in Fig.8, based on the multi-carrier algorithm, the input data is first coded by Low Density Parity Check Codes (LDPC). After baseband mapping, the data sequence is converted to parallel and enters the first IFFT to perform CI spreading. Pilot signals are inserted before the second IFFT. The second IFFT is used to implement orthogonal multi-carrier modulation. A cyclic prefix and postfix are also appended to the data sequence as guard intervals in order to combat the ISI induced by the multi-path delay spread in the UWA channel. The complex base-band signal is then up-converted to the transmission frequency and the real part of the signal is sent out to the UWA channel by the transducer.

In the receiver, the signal is first down-converted to the base-band. Then the signal is demodulated by the first FFT. Channel estimation is performed to track the channel response and compensations of the signal are performed. Then CI code de-spreading is implemented by the second FFT, and finally, the phase constellation of the data is extracted.

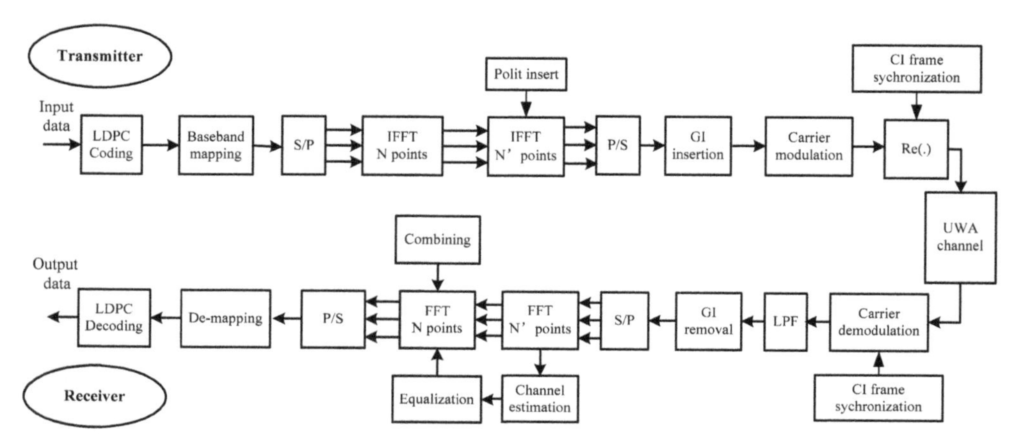

Fig. 8. The block diagram of the system based on multi-carrier algorithm

Fig.9 shows a simplified block diagram of the proposed system based on single-carrier algorithm. In the transmitter, the input data is first encoded by LDPC and then mapping into a baseband constellation. The data sequence is converted to parallel and enters the first FFT to perform CI spreading. After that, the IFFT is used to implement orthogonal multi-carrier modulation. A cyclic prefix is also appended as a guard interval to the data sequence in order to combat the inter ISI induced by multipath delay spread in the selective fading channel. In addition, a pilot signal is appended for the purposes of channel estimation in the receiver. The complex baseband signal is then up-converted to the transmission frequency. In the receiver, the signal is down-converted to baseband. The signal is first demodulated by FFT, and diversity combining scheme is employed as frequency-domain equalization where the combining weights are estimated by the pilot signal. Then de-spreading is implemented by the IFFT and the phase constellation of the data is extracted. Finally, the data is mapped back to the binary form, and a soft LDPC decoding is performed.

We here focus on the multi-carrier algorithm and explain the key technologies used in the underwater acoustic communication system.

4.1 Synchronization

We use Linear Frequency Modulation (LFM) (Rihaczek, 1969; Shaw & Srivastava, 2007) signal to get coarse synchronization and CI complex spreading sequence to get accurate synchronization and fractional frequency offset estimation.

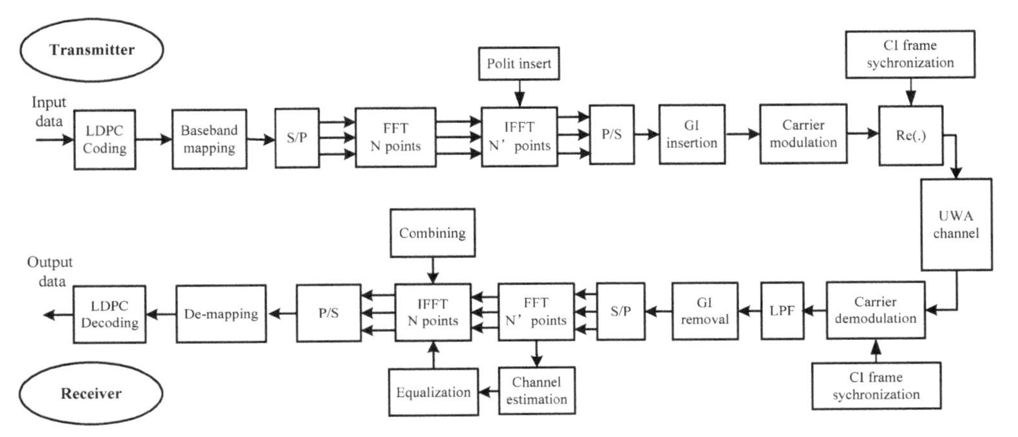

Fig. 9. The block diagram of the system based on single-carrier algorithm

4.1.1 Coarse synchronization

The expression for LFM signal is given as

$$s(t) = A \cdot rect(\frac{t}{T})e^{j2\pi\left(f_0 t + \frac{1}{2}ut^2\right)} \tag{16}$$

where A is the amplitude of the signal, T is the width of the signal, f_0 is the carrier frequency, u is the gradient of the instantaneous frequency which is called chirp rate, $u = 2\pi B/T$. $rect(\cdot)$ is a rectangle function, defined as

$$rect(\frac{t}{T}) = \begin{cases} 1, & \left|\frac{t}{T}\right| \le \frac{1}{2} \\ 0, & \left|\frac{t}{T}\right| > \frac{1}{2} \end{cases} \tag{17}$$

Since the ambiguity function of LFM signals is wide in Doppler axis (Rihaczek, 1969), it is highly tolerant of the Doppler shift which makes it useful in mobile wireless communication systems.

4.1.2 Fine synchronization and fractional frequency shift estimation

Two identical CI complex sequences are used as fine synchronization signals and sliding correlator is applied at the receiver side to obtain the correlation peak. CI complex sequences is given by

$$1,1,...,1,1,\xi_1,\xi_2,\xi_3,....,\xi_{N-1},1,\xi_1^2,\xi_2^2,...,\xi_{N-1}^2,1,\xi_1^3,\xi_2^3,...,\xi_{N-1}^3,1,\xi_1^{N-1},\xi_2^{N-1},...,\xi_{N-1}^{N-1} \quad (18)$$

where $\xi_k = e^{j(2\pi k/N)}, 0 \le k \le N-1$. Suppose the two received sequences are $r_1(m)$ and $r_2(m)$ (Hlaing et al. 2003; Ren, 2005), the cross-correlation function of the two sequences is

$$R(n) = \sum_{m=0}^{L-1} (r_1^*(m) r_2(m+n)) \quad (19)$$

where the L is the length of the sequence. Since the sequences are identical at the transmitter, the impact of the channel is assumed to be same to the two sequences, (18) can be written as

$$R(n) = \sum_{m=0}^{L-1} \left(r_1^*(m) r_2(m+n) \right) = \sum_{m=0}^{L-1} \left(r_1^*(m) r_1(m+L+n) \right) \quad (20)$$

The cross-correlation of two sequences can be changed into auto-correlation of one sequence, that is

$$R(0) = \sum_{m=0}^{L-1} \left(r_1^*(m) r_2(m) \right) = \sum_{m=0}^{L-1} \left(r_1^*(m) r_1(m+L) \right) \quad (21)$$

The time offset can be estimated from

$$\hat{n} = \arg\max_n \left(|R(n)|^2 \right) \quad (22)$$

Assuming that the frame synchronization is accurate, the difference between two CI complex sequences can be approximately regarded as the result of the frequency shift

$$\theta = angle(R(\hat{n})) \quad (23)$$

$$\Delta f = \theta / (2\pi T/2) = \theta / (\pi T) \quad (24)$$

where Δf is the fractional frequency shift, θ is the phase offset caused by frequency shift, T is the period of the synchronization signal which is equals to the CI/OFDM signal.

Fig. 10 shows the sliding correlation peaks of LFM signal and CI complex sequence at the receiver side in AWGN channel.

4.2 Channel estimation and equalization

When the pilot is a CAZAC sequence, it was proven that in the presence of noise, the mean square error of the channel response estimation is minus (Milewski, 1983). The mean square error equals to the variance of the noise in the channel, that is

$$E(\sum_i |\hat{r}_i - r_i|^2) = L\sigma^2 \sum_i |v_i|^2 = \sigma^2 \quad (25)$$

where $v_i \in V$, $V = 1/U$, U is the Fourier transform of the pilot sequence.

Two kinds of pilot sequences which are both CAZAC sequences are chosen to estimate the underwater acoustic channel response.

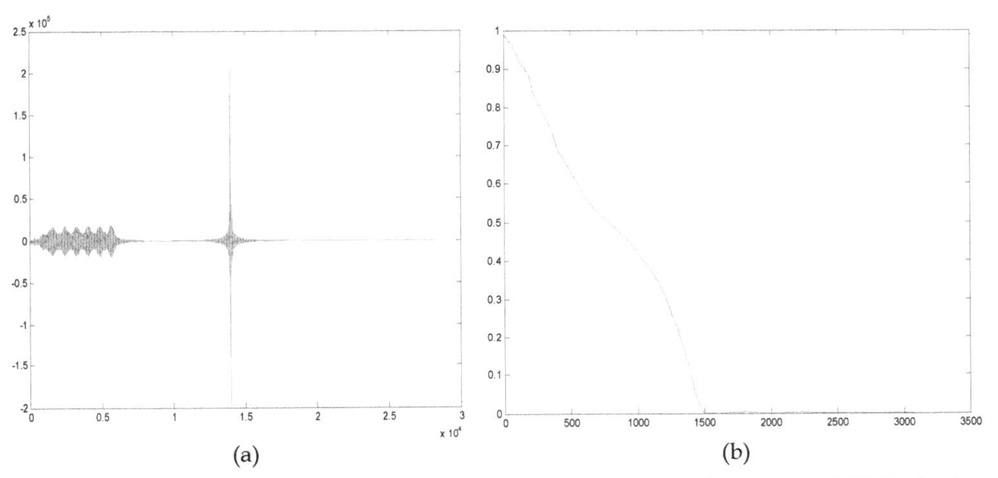

Fig. 10. Sliding correlation peak under AWGN channel. (a) Peak of LFM signal (b) Peak of CI spreading sequence

4.2.1 Pilot

1. CI complex sequence

CI complex sequence was given in equation (18). It is easy to prove the CAZAC feature of the CI complex sequence. Fig. 11 gives simulation results of the CI complex sequence.

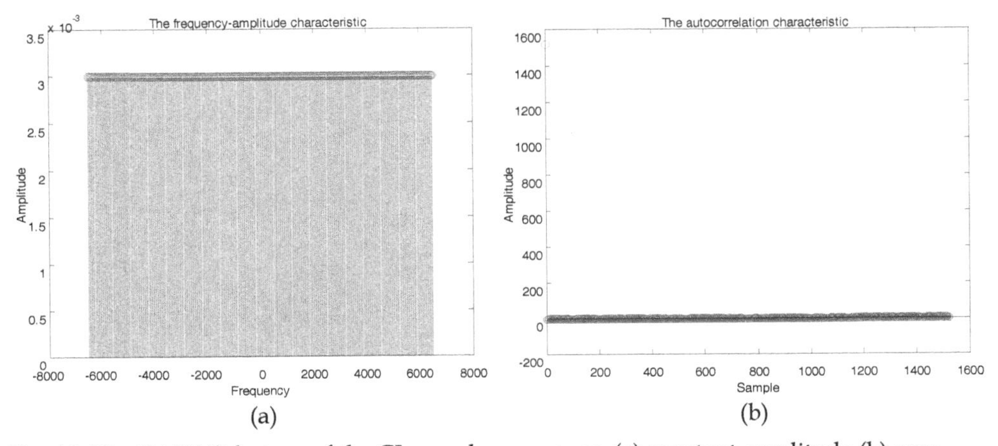

Fig. 11. The CAZAC feature of the CI complex sequence. (a) constant amplitude (b) zero autocorrelation

2. CHU sequence

CHU sequence is a polyphase code with a periodic autocorrelation function (CHU, 1972). It was proven that the CHU sequence can be constructed for any code length. When N is even, CHU sequence is defined as $a_k = \exp(i\frac{M\pi k^2}{N})$. When N is odd, it is $a_k = \exp(i\frac{M\pi k(k+1)}{N})$, where M is an integer relatively prime to N. Fig. 12 shows the amplitude-frequency and autocorrelation results of CHU sequence. Fig. 13 and Fig.14 show the channel impulse respond estimated by CI complex sequence and CHU sequence under AWGN and 4-path Rayleigh channel.

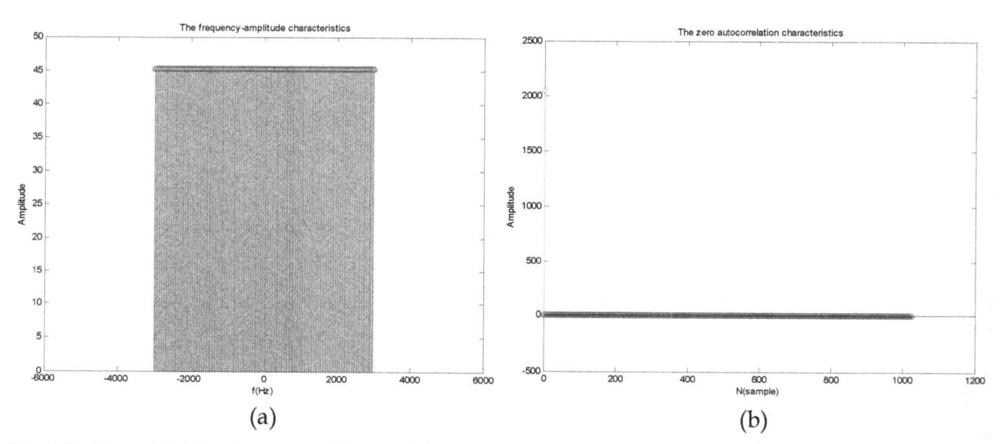

(a) (b)

Fig. 12. The CAZAC feature of the CHU sequence (a) constant amplitude (b) zero autocorrelation

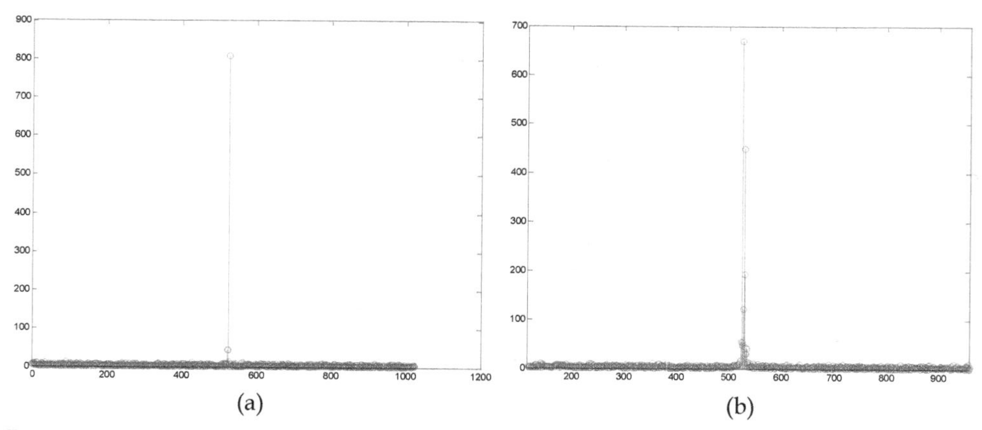

(a) (b)

Fig. 13. Channel impulse respond estimated by CI complex sequence (a) AWGN channel (b) 4-path Rayleigh channel

4.2.2 Frequency domain equalization

In the conceptual model of CI/OFDM, the information symbols are simultaneously modulated on multi-subcarrier which makes CI/OFDM inherently having the frequency diversity. At the receiver side, frequency combining may be used to improve the performance of the system.

We still focus on the multi-carrier algorithm which makes use of the properties of IDFT/DFT, such as the linearity and the circular shift. Since every parallel input signal of the second IDFT is the summation of information signals, which are spread by CI signal, the frequency diversity combining should be at the end of the first DFT module at the receiver side.At the receiver, after orthogonal multi-carrier demodulation, the output of the first FFT is

$$r_i = H_i \cdot \sum_{k=0}^{N-1} a_k e^{j\frac{2\pi}{N} \cdot i \cdot k} + n_i \tag{26}$$

where $i \in [1,N]$ is the number of the subcarrier, H_i is the transition function of the sub-channel and a_k is the information symbol, $k \in [1,N]$.

According to Fig. 15, the input signals at the second FFT module which can be expressed as

$$R_i = w_i r_i = w_i (H_i \sum_{k=0}^{N-1} a_k e^{j\frac{2\pi}{N} \cdot i \cdot k} + n_i) \tag{27}$$

where w_i is the gain of the ith subcarrier determined by different combining strategies. Since R_i includes all information about transmitted symbols, the second FFT not only is applied for decomposing the CI spreading into the subcarrier components, but also is used to complete the frequency combining.

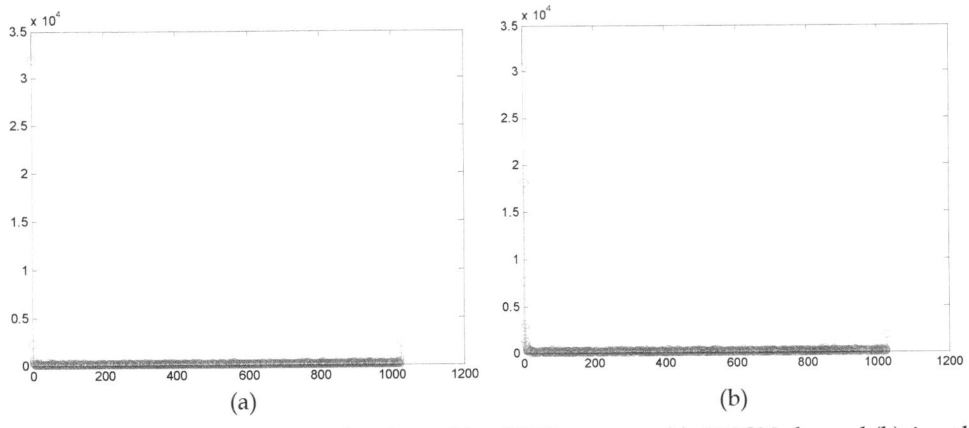

Fig. 14. Channel impulse respond estimated by CHU sequence (a) AWGN channel (b) 4-path Rayleigh channel

Several frequency combining strategies such as EGC, the MRC, orthogonal restoration combining (ORC) and MMSEC are considered in frequency selective channel (Itagkai & Adachi, 2004). Computer simulation results are shown in Fig. 16. In CI/OFDM system, though both combing strategies are sensitive to the inter-carrier interference (ICI), the MRC performance is worse than EGC. It is because that in MRC, the phase changes of the signals are lost, which is very important to CI/OFDM. Since ORC equalization perfectly restores frequency non-selective channel but produces the noise enhancement, its performance is better than MRC and EGC but worse than MMSEC. As SNR increases, the BER performance of ORC gets better, but MMSEC equalization provides the best performance. AS we know, the MMSEC not only restores frequency non-selective channel but also minimizes the noise.

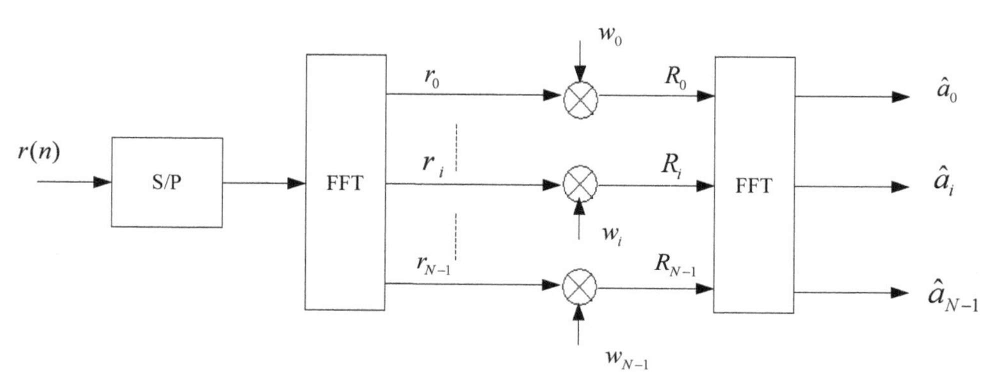

Fig. 15. Frequency combining in multi-carrier algorithm

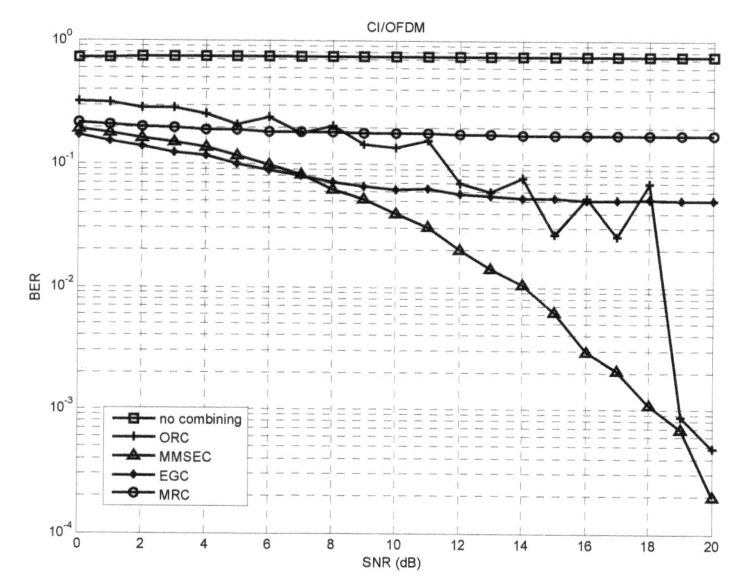

Fig. 16. Performance comparisons of ORC, MMSEC, EGC and MRC

5. Field test results

5.1 Experiment I

5.1.1 Experimental pool

The experimental pool in Xiamen University is an un-censored pool which size is 430(L)x320(W)x200(H)cm.

Channel estimation was carried out, 13kHz single carrier signal is sent at the 30ms intervals. The sample rate of A/D is 160kHz. The transmitted and received signals are shown in Fig. 17. We can see that key features of the channel are multi-path transmission with low noise because of the stationary water in the pool. The maximum delay is about 19 ms which magnitude is under 3% of the maximum one.

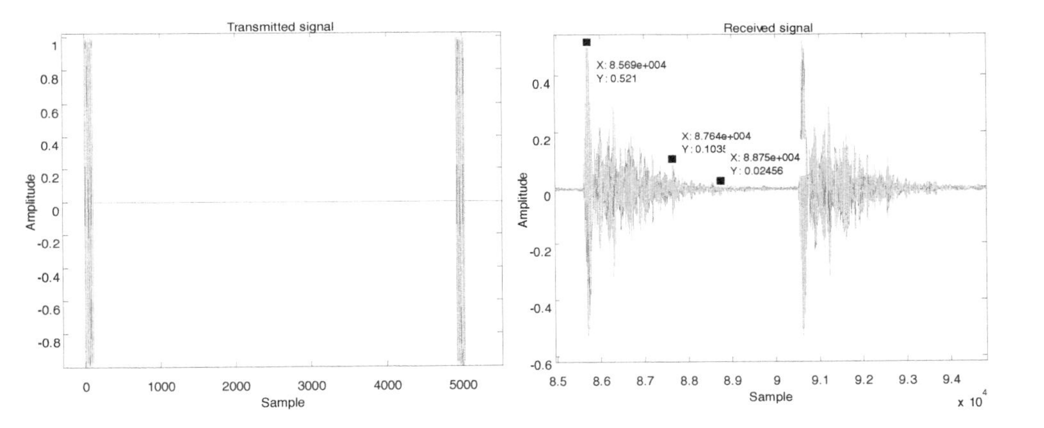

Fig. 17. The transmitted and received signal (13kHz)

5.1.2 Results of experimental pool

In this experiment, LFM signal was used to get coarse synchronization of the frame and CI complex sequence was applied to achieve accurate synchronization and fractional frequency shift estimation (Xu et al., 2008). SNR estimation algorithm is borrowed from the work (Ren, 2005) with synthesis of CI complex sequence. CI complex sequence is used as pilot to estimate the impulse response of the channel. Frequency-domain equalization ORC is used at the receiver to improve the performance of the system. Table 1 shows the parameters of the CI/OFDM system.

The results are shown in Table 2. Since the water in pool is stationary, the SNR is high and the average SNR is about 12dB. BER performance is good and the average fractional frequency shift is about 0.07Hz, which is very small compared with the subcarrier interval $\Delta f = 6000 / 1024 = 5.86Hz$.

Baseband mapping	QPSK
Subcarrier mapping	Localized
Synchronization signal	LFM/CI complex spreading sequence
Pilot	CI complex sequence
Pilot pattern	Block
Equalization	ORC
Bandwidth	6kHz
Carrier frequency	13kHz
Sampling rate	156kHz
Number of the parallel signal	1024
System rate	4.97kbps

Table 1. System parameters

SNR (dB)	Fractional frequency shift (Hz)	BER
12.33659	0.068643	3.48771e-06

Table 2. BER performance

5.2 Experiment II

5.2.1 Shallow water in Wuyuan Bay of Xiamen

A CI/OFDM underwater acoustic communication experiment was conducted on Dec. 12, 2008 in Wuyuan Bay of Xiamen, China. The distance between the transmitter and the receiver was 1000m, and they were deployed at 3 m and 2 m below the sea-surface, respectively. The average depth of the water is 4. 5m which was changed with the tide. The channel probing signals include two kinds. One is a single carrier signal which was transmitted at the 100ms intervals repeatedly. The other is a sweeping signal which frequency is from 9kHz to 21kHz. It was also be transmitted once every 100ms.

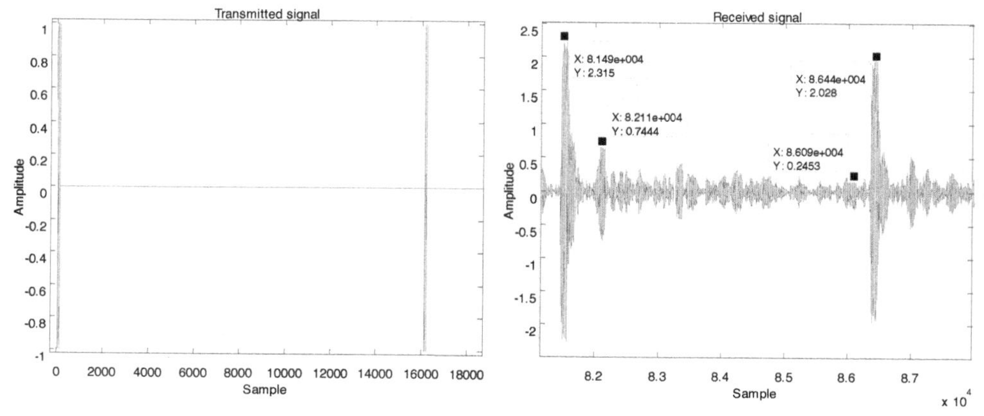

Fig. 18. Transmitted and received signal (14kHz)

As shown in Fig. 18, multi-path signals is visible during the 100ms intervals though the amplitude of the multi-path signal is about 2% of the maximum one. The amplitude of the 14KHz signal was changed after 100ms. Fig. 19 denotes the different amplitudes of the sweeping signals which reveal the time-varying and frequency-varying features of the acoustic underwater channel.

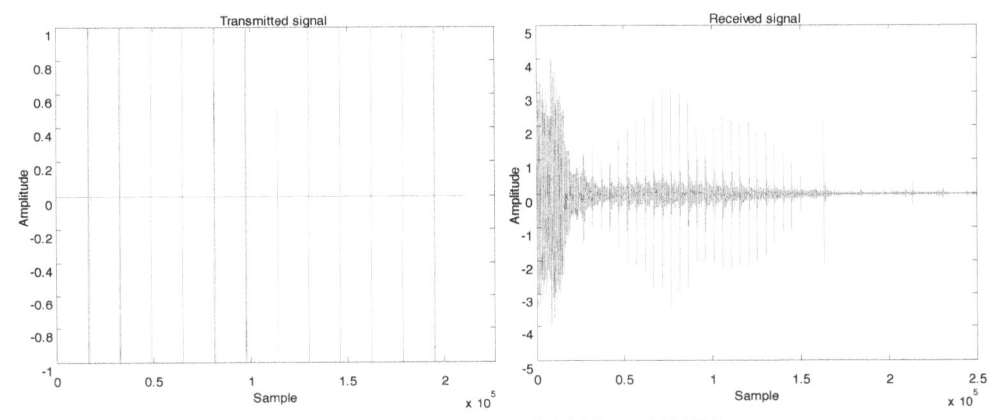

Fig. 19. Transmitted and received sweeping signal (9kHz to 21kHz)

5.2.2 Results of experiment II

Table 3 is the parameters of CI/OFDM underwater acoustic communication system. Table 4 is the BER performance of the system without and with frequency selective combining. The BER performance was significantly improved due to the frequency diversity combining by the cost of decrease of the data rate.

Baseband mapping	QPSK
Subcarrier mapping	Localized
Synchronization signal	LFM
Pilot	CI/OFDM signal
Pilot pattern	Block
Equalization	MMSEC
Bandwidth	5kHz
Carrier frequency	15kHz
Sampling rate	60kHz
Number of the parallel signal	1024
Data rate	4.97kbps/1.24kbps in 4-fold frequency diversity

Table 3. System parameters

BER (without frequency diversity)	BER (with 4-fold frequency diversity)
0.014257	0.0058714

Table 4. BER performance

5.3 Experiment III

5.3.1 Shallow water in Baicheng water

Two CI/OFDM underwater acoustic communication experiments were conducted on Dec. 17, 2009 and Dec. 31, 2009 in Baicheng water of Xiamen, China, respectively. The transmitter was deployed at 2.5m above the sea-floor in 5.5m deep water and the receiver was deployed at 9m below the sea-surface in 16m deep water. The horizontal distance between the transmitter and receiver were 2000m and 5000m, respectively.

The channel probing signals used in these two experiments were same sweeping signals, with frequency from 8KHz to 16KHz. The time interval between different frequencies was 30 ms.

Fig.20 is the received probing signal at the short range (2000m). The sea condition was calm but vessels passed through the water frequently. The Frequency-varying feature is different from the feature in Wuyuan Bay. In this experiment, the amplitudes of the lower and higher frequency were faded significantly. The ambient noise was much higher in this underwater channel. From the enlarged map of received signal of 14KHz and 14. 5KHz, it is easy to see that the amplitudes of the strongest arrival changed with time and frequencies. The high level of noise made it difficult to distinguish multi-path signals from the ambient noise. Note that there is no apparent impulse interference and the amplitudes of multi-path signals are much smaller than that of the main path.

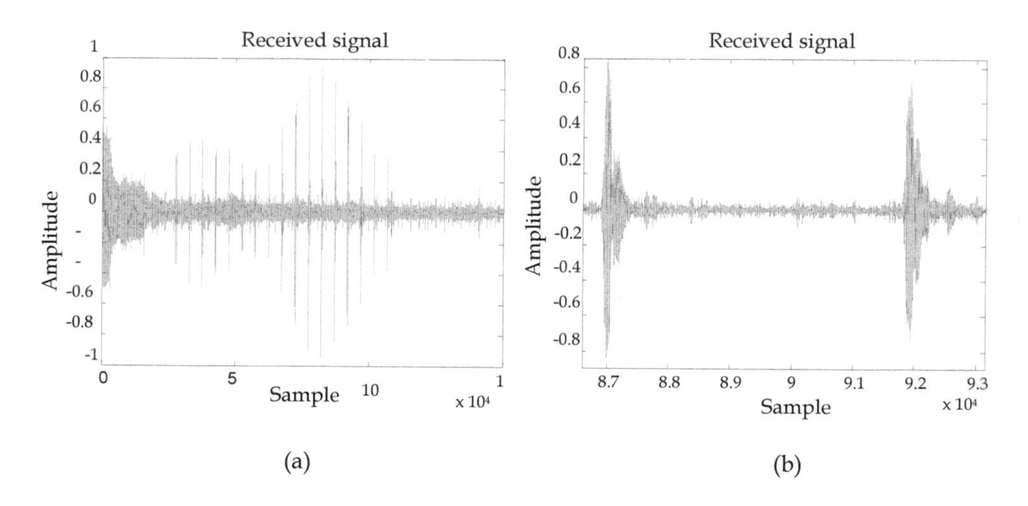

Fig. 20. Received signals. (a) received sweeping signals (8kHz to 16kHz) (b) received single carrier signal (14kHz and 14.5kHz)

Fig. 21. is the transmitted and received probing signal at the long range (5000m). It was a windy day, and the sea condition was not calm. Many vessels passed through the water.

The Frequency-varying feature is different from the feature in short range at the same water domain. In this experiment, the ambient noise was much higher than that in the short range.

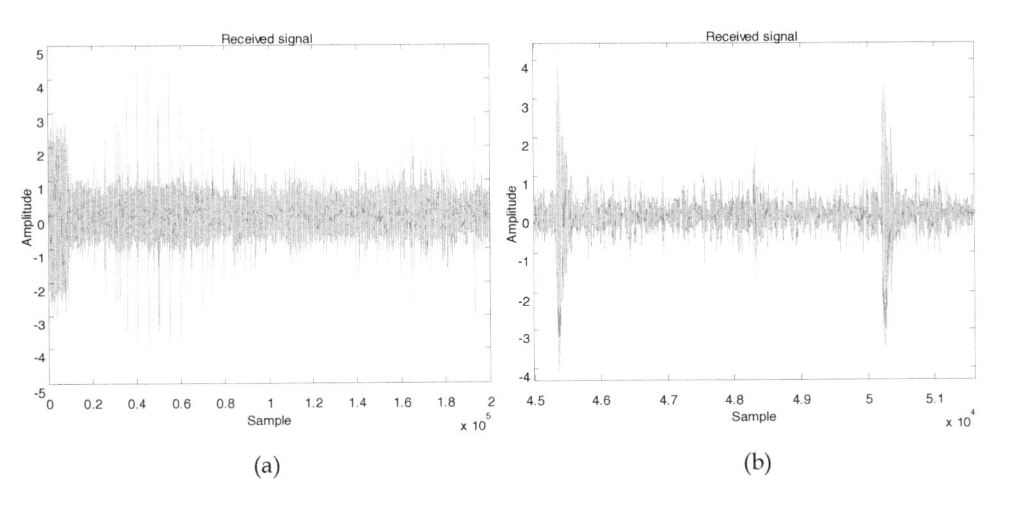

Fig. 21. Received signals. (a) received sweeping signals (8kHz to 16kHz) (b) received single carrier signal (14kHz and 14.5kHz)

5.3.2 Results of experiment III

Based on the channel probing results, we concluded that the channel in Baicheng water was worse than that in Wuyuan Bay. In experiments, 4-fold frequency diversity and (2,1) LDPC were applied in CI/OFDM systems (Bai et al., 2009) in order to guarantee the BER performance of the system.

System parameters are same in Table 3 except that the comb pilot pattern is used instead of the block pilot. The results of using (2,1) LDPC is the performance improvement and the decrease of the data rate.

Date	BER (before LDPC decoding)	BER (after LDPC decoding)
Dec. 17, 2009	0.0393	0
Dec. 31, 2009	0.06598	0

Table 5. BER performance

Fig. 22 is the received CI/OFDM signals in short range (2000m) experiment. The amplitude of the received signal changed dramatically, and the level of ambient noise was high. There was deeply frequency fading in the bandwidth of the signal. It might explain the reason of BER performance degradation even though the 4-fold frequency diversities were used.

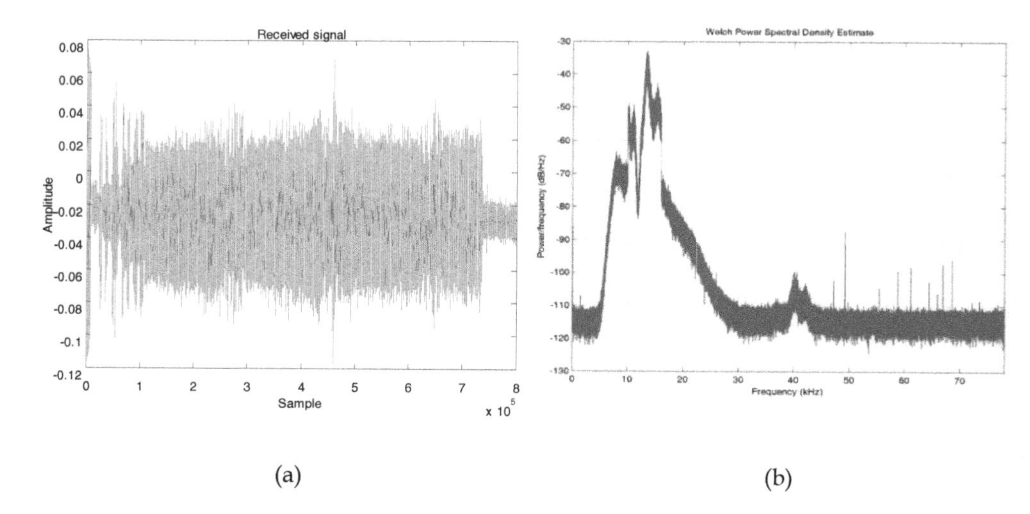

Fig. 22. The received signal (a) the received profile (b) the power spectral density of the received signal

6. Conclusion

In this chapter, we first proposed two algorithms which simplify the modulation and demodulation of the conceptual CI/OFDM. Secondly, based on these algorithms and jointed with synchronization, channel estimation and equalization, we constructed CI/OFDM underwater acoustic communication systems. In the end, a number of experiments were carried out in the experimental pool and shallow waters in Xiamen of China to verify the performance of the system. Field results are as followed:

The BER of the uncoded CI/OFDM underwater acoustic communication system at the data rate 4.97kbps is lower than 4×10^{-6} in the experimental pool (experiment I) and 1.5×10^{-2} in Wuyuan Bay in Xiamen (experiment II). When 4-fold frequency diversity is applied, the data rate is 1.24kbps and the BER performance of the system is lower than 6×10^{-3} in experiment II.

The BER of the uncoded 4-fold frequency diversity CI/OFDM acoustic communication system at the data rate 1.24kbps is lower than 4×10^{-2} and 7×10^{-2} in experiment III at Baicheng water in Xiamen. The BER of the coded frequency diversity CI/OFDM acoustic communication system at the data rate 620bps is almost zero in the short range and the long range.

A few problems exist in the system. The most important one is that the inherent frequency diversity of CI/OFDM did not play its due role in the system. According to our analysis, the orthogonality between the different symbols modulated on the same subcarrier will be destroyed if the phases were changed when signals transmitted in the channel. It means that the intersymbol interference cancels out the diversity combining gain. Future researches should focus on the optimization of the algorithms in order to take advantage of the inherent frequency diversity and other realizations based on the conceptual CI/OFDM.

7. Acknowledgment

This work was supported by the Fundamental Research Funds for the Central Universities of China 2010121062.

8. References

Bai, L. Y., Xu, F., Xu, R & Zheng S. Y. (2009), LDPC Application Based on CI/OFDM Underwater Acoustic Communication System, *Proceedings of the First International Conference on Information Science and Engineering*, ISBN 978-0-7695-3887-7, Nanjing, Jiangsu China, December 26-December 28, 2009

Berkhovskikh, L. & Lysanov, Y. (2003). *Fundamentals of Ocean Acoustics* (3rd edition), Springer, ISBN 0-387-95467-8

CHU, D. C. (1972), Polyphase codes with good periodic correlation properties, *IEEE Transactions on information theory*, Vol. 18, Issue 4, pp. 531 - 532, July, 1972, ISSN 0018-9448

Freitag, L., Stojanovic, M, Singh, S. & Johnson M. (2001), Analysis of channel effects on direct-sequence and frequency-hopped spread-spectrum acoustic communication, *IEEE Journal of Oceanic Engineering*, Vol. 26, Issue 4, pp. 586-593, Oct 2001, ISSN 0364-9059

Frassati, F., Lafon, C., Laurent, P. A. & Passerieux, J. M. (2005), Experimental assessment of OFDM and DSSS modulations for use in littoral waters underwater acoustic communications, *Proceedings of Oceans 2005, Europe*, 20-23 June, 2005

Heimiller, R. C. (1961), Phase shift pulse codes with good periodic correlation properties, *IRE Transactions on Information Theory*, Vol. 7, Issue 4, pp. 254 - 257, October 1961, ISSN 0096-1000

Hlaing, M., Bhargava, V. K. & Letaief, K. B. (2003), A robust timing and frequency synchronization for OFDM systems, *IEEE Transactions on Communications*, Vol. 2, Issue 4, pp. 822 - 839, July 2003, ISSN 1536-1276

Itagkai, T. & Adachi F. (2004), Joint frequency domain equalization and antenna diversity combining for orthogonal multicode DS-CDMA signal transmissions in a frequency selective fading channel, *IEICE Transactions on Communications*, Vol. E87-B, No. 7, pp. 1954-1963, July 2004, ISSN 0916-8516

Jensen, F., Kuperman, W., Porter, M. & Schmidt H. (2011). *Computational Ocean Acoustics* (2nded), Springer,. ISBN 978-1441986771

Kim, B. C. & Lu, I. T. (2000), Parameter study of OFDM underwater communications system, *Proceedings of MTS/IEEE Conference and Exhibition OCEANS' 2000*, Providence, RI , USA, 11-14 Sept., 2000

Lam, W. K. and Ormondroyd R. F. (1997), A coherent COFDM modulation system for a time-varying frequency-selective underwater acoustic channel, *Proceedings of Seventh International Conference on Technology Transfer from Research to Industry, Electronic Engineering in Oceanography*, 23-25 June, 1997

Li, B. S., Zhou, S. L., Stojanovic, M., Freitag, L. & Willett, P. (2008), Multi-carrier communication over underwater acoustic channels with nonuniform Doppler shifts, *IEEE Journal of Oceanic Engineering*, Vol. 33, Issue 2, pp. 198-209, April 2008, I SSN: 0364-9059

Milewski, A. (1983), Periodic sequences with optimal properties for channel estimation and fast start-up equalization, *IBM Journal of Research and Development*, Vol. 27, Issue 5, pp. 426-431, Sept. 1983, ISSN 0018-8646

Nassar, C. R.; Natarajan, B. & Shattil, S. (1999). Introduction of carrier interference to spread spectrum multiple access, IEEE Emerging Technologies Symposium, Dallas, Texas,Apr 12-13, 1999

Nassar, C. R., Natarajan, B., Wu, Z., Wiegandt,D. A.; Zekavat S. A. & Shattil, S. (2002) *High-performance, Multi-carrier technologies for wireless communications*, Kluwer Academic Publishers, ISBN 0792-347618-8

Nassar, C. R., Natarajan, B., Wiegandt,D. A. & Wu, Z. (2002), Multi –carrier platform for wireless communications. Part 2: OFDM and MC-CDMA systems with high performance, high throughput via innovations in spreading, *Jornal of Wireless Communications and Mobile Computing*, Vol. 2, Issue 4, 2002. pp. 381-403, ISSN 1530-8677

Ren G. L., Chang Y. L., Zhang H, & Zhang H. N. (2005), Synchronization Method Based on a New Constant Envelop Preamble for OFDM Systems, *IEEE Transactions on Broadcasting*, Vol. 51, NO. 1, pp. 139-143, March 2005, ISSN 0018-9316

Rihaczek, A. W. (1977), *Principles of high-resolution radar*, Mark Resources, Inc. ISBN 978-0890069004

Shaw, A. and Srivastava, S. (2007), A Novel preamble structure for robust timing synchronization in OFDM system, *Proceedings of IEEE Region 10 Conference on TENCON*, ISBN 978-1-4244-1272-3, Taipei, China, Oct. 30 2007-Nov. 2 2007

Stojanovic, M. (1996), Recent advances in high-speed underwater acoustic communications, *IEEE Jounal of Oceanic Engineering*, Vo. 21, Issue 2, pp. 125–136, Apr. 1996. ISSN 0364-9059

Stojanovi, M., Proakis, J. G, Rice J. A. & Green, M.D.(1998), Spread spectrum underwater acoustic telemetry, Proceedings of Oceans 98, Nice, France Sep28 - Oct 1, 1998

Stojanovic & M. Freitag, L (2006), Multichannel Detection for Wideband Underwater Acoustic CDMA Communications, *IEEE Journal of Oceanic Engineering*, Vol. 31, Issue 3, pp. 685-695, July 2006, ISSN 0364-9059

Stojanovic, M. (2006), Low complexity OFDM detector for underwater acoustic channels, *Proceedings of MTS/IEEE Conference and Exhibition OCEANS' 2006,* Boston, Massachusetts, USA, 18-21 Sept. 2006

Stojanovic, M. (2007). On the Relationship Between Capacity and Distance in an Underwater Acoustic Channel, *ACM SIGMOBILE - Mobile Computing and Communications Review,* vol. 11, No. 4, Oct. 2007, pp. 34–43, ISSN: 1559-1662

Stojanovic, M. (2008), OFDM for underwater acoustic communications: adaptive synchronization and sparse channel estimation, *Proceedings of IEEE International Conference on Acoustics, Speech and Signal Processing,* ISSN 1520-6149, Las Vegas, NV, March 31 2008-April 4, 2008

Stojanovic, M. & Preisig, J. (2009). Underwater Acoustic Communication Channels: Propagation Models and Statistical Characterization, *IEEE Communications Magazine,* Vol. 47, Issue 1, Jan. 2009, pp. 84-89, ISSN 0631-6804

Tsimenidis, C. C., Hinton, O. R., Adams, A. E. & Sharif, B. S. (2001), Underwater acoustic receiver employing direct-sequence spread spectrum and spatial diversity combining for shallow-water multiaccess networking, *IEEE Journal of Oceanic Engineering,* Vol. 26, Issue 4, pp. 594-603, Oct 2001, ISSN 0364-9059

Wiegandt,D. A. & Nassar, C. R. (2001). High Performance OFDM via carrier interferometry, *Proceedings of IEEE International Conference on Third Generation Wireless and Beyond (3Gwireless '01),* San Francisco, CA, May 30, 2001

Wiegandt,D. A.&Nassar, C. R. (2001), "Crest Factor Considerations in MC-CDMA with Carrier Interferometry Codes," *Proceedings of IEEE Pacific-Rim Conference on Communications,Computers and Signal Processing,* Victoria, Canada, August 26-28, 2001

Wiegandt,D. A.; Nassar, C. R. &Wu Z. (2001). Overcoming peak-to-average power ratio issues in OFDM via carrier interferometry codes, *Proceedings of IEEE International Conference on Vehicular Technology,* Atlantic City, NJ, USA, Oct. 7-11, 2001.

Wiegandt,D. A.&Nassar, C. R. (2003). High-throughput, high-performance OFDM via pseudo-orthogonal carrier interferometry spreading codes. *IEEE transaction of Communications,* Vol. 51, Issue 7, July 2003, pp. 1123-1134, ISSN 0090-6778

Wiegandt,D. A.; Nassar, C. R. &Wu Z. (2004). The elimination of peak-to-average power ratio concerns in OFDM via carrier interferometry spreading codes: a multiple constellation analysis, *Proceedings of the Thirty-Sixth Southeastern Symposium on System Theory,* March 14-16, 2004

Xu, F., Hu, X.Y. & Xu, R. (2007), A Novel Implementation of Carrier Interferometry OFDM in an Underwater Acoustic Channel, *Proceedings of Oceans 2007,* ISBN 978-1-4244-0635-7, Aberdeen, *UK,* 18-21 June 2007

Xu, F., Xu, R. & Sun H. X. (2007), Implementation of Carrier Interferometry OFDM by Using Pulse Shaping Technique in Frequency Domain, *Proceedings of IEEE International Workshop on Anti-counterfeiting, Security, Identification,* ISBN 1-4244-1035-5, Xiamen, Fujian, China, 16-18 April, 2007

Xu, F. Xu, R., Zhang Y. H., Sun H. X. & Wang, D. Q. (2008), Design and test of a carrier interferometry OFDM system in underwater acoustic channels, *Proceedings of IEEE International Conference on Communications, Circuits and Systems*, ISBN 978-1-4244-2063-6, Fujian, China, 25-27 May, 2008

Permissions

The contributors of this book come from diverse backgrounds, making this book a truly international effort. This book will bring forth new frontiers with its revolutionizing research information and detailed analysis of the nascent developments around the world.

We would like to thank Salah Bourennane, for lending his expertise to make the book truly unique. He has played a crucial role in the development of this book. Without his invaluable contribution this book wouldn't have been possible. He has made vital efforts to compile up to date information on the varied aspects of this subject to make this book a valuable addition to the collection of many professionals and students.

This book was conceptualized with the vision of imparting up-to-date information and advanced data in this field. To ensure the same, a matchless editorial board was set up. Every individual on the board went through rigorous rounds of assessment to prove their worth. After which they invested a large part of their time researching and compiling the most relevant data for our readers. Conferences and sessions were held from time to time between the editorial board and the contributing authors to present the data in the most comprehensible form. The editorial team has worked tirelessly to provide valuable and valid information to help people across the globe.

Every chapter published in this book has been scrutinized by our experts. Their significance has been extensively debated. The topics covered herein carry significant findings which will fuel the growth of the discipline. They may even be implemented as practical applications or may be referred to as a beginning point for another development.

The editorial board has been involved in producing this book since its inception. They have spent rigorous hours researching and exploring the diverse topics which have resulted in the successful publishing of this book. They have passed on their knowledge of decades through this book. To expedite this challenging task, the publisher supported the team at every step. A small team of assistant editors was also appointed to further simplify the editing procedure and attain best results for the readers.

Our editorial team has been hand-picked from every corner of the world. Their multi-ethnicity adds dynamic inputs to the discussions which result in innovative outcomes. These outcomes are then further discussed with the researchers and contributors who give their valuable feedback and opinion regarding the same. The feedback is then collaborated with the researches and they are edited in a comprehensive manner to aid the understanding of the subject.

Apart from the editorial board, the designing team has also invested a significant amount of their time in understanding the subject and creating the most relevant covers. They scrutinized every image to scout for the most suitable representation of the subject and create an appropriate cover for the book.

The publishing team has been involved in this book since its early stages. They were actively engaged in every process, be it collecting the data, connecting with the contributors or procuring relevant information. The team has been an ardent support to the editorial, designing and production team. Their endless efforts to recruit the best for this project, has resulted in the accomplishment of this book. They are a veteran in the field of academics and their pool of knowledge is as vast as their experience in printing. Their expertise and guidance has proved useful at every step. Their uncompromising quality standards have made this book an exceptional effort. Their encouragement from time to time has been an inspiration for everyone.

The publisher and the editorial board hope that this book will prove to be a valuable piece of knowledge for researchers, students, practitioners and scholars across the globe.

List of Contributors

Said Assous and Mike Lovell
Ultrasound Research Laboratory, University of Leicester, United Kingdom

Laurie Linnett
Fortkey Ltd, United Kingdom

David Gunn, Peter Jackson and John Rees
Ultrasound Research Laboratory, British Geological Survey, United Kingdom

Caroline Fossati, Salah Bourennane and Julien Marot
Institut Fresnel, Ecole Centrale Marseille, France

Shen Xiaohong, Wang Haiyan, Zhang Yuzhi and Zhao Ruiqin
College of Marine Engineering, Northwestern Polytechnical University, Xi'an, China

Salah Bourennane, Caroline Fossati and Julien Marot
Institute Fresnel, Ecole Centrale Marseille, France

Weijie Shen, Haixin Sun, En Cheng, Wei Su and Yonghuai Zhang
The Key Laboratory of Underwater Acoustic Communication and Marine Information Technology, Xiamen University, Xiamen, Fujian, China

Liang Zhao and Jianhua Ge
State Key Laboratory of Integrated Service Networks, Xidian University, Xi'an, P.R. China

Fang Xu and Ru Xu
Xiamen University, China